ENCOUNTERING ETI

Books by John Hart

The Spirit of the Earth: A Theology of the Land (1984)

Ethics and Technology: Innovation and Transformation in Community Contexts (1997)

What Are They Saying about Environmental Theology? (2004)

Sacramental Commons: Christian Ecological Ethics (2006)

Cosmic Commons: Spirit, Science, and Space (2013)

ENCOUNTERING ETI

Aliens in *Avatar* and the Americas

John Hart

CASCADE *Books* • Eugene, Oregon

ENCOUNTERING ETI
Aliens in *Avatar* and the Americas

Copyright © 2014 John Hart. All rights reserved. Except for brief quotations in critical publications or reviews, no part of this book may be reproduced in any manner without prior written permission from the publisher. Write: Permissions. Wipf and Stock Publishers, 199 W. 8th Ave., Suite 3, Eugene, OR 97401.

Cascade Books
An Imprint of Wipf and Stock Publishers
199 W. 8th Ave., Suite 3
Eugene, OR 97401

www.wipfandstock.com

ISBN 13: 978-1-61097-880-4

Cataloguing-in-Publication Data

Hart, John, 1943–

    Encountering ETI: aliens in *Avatar* and the Americas / John Hart.

    x + 296 p. ; 23 cm.

    ISBN 13: 978-1-61097-880-4

    1. Unidentified flying objects—Religious aspects. 2. Space theology. I. Title.

BL65.U54 H36 2014

Manufactured in the U.S.A.   10/28/2014

The Scripture quotations contained herein are from the New Revised Standard Version Bible, copyright © 1989, Division of Christian Education of the National Council of Churches of Christ in the U.S.A. Used by permission. All rights reserved.

To those who have a cosmic consciousness: who explore creatively, consider thoughtfully, and contemplate responsibly the unexpected realities they encounter, and who conscientiously share their discoveries in order to promote the socioecological wellbeing of Earth and all biotic being;

To astrophysicist J. Allen Hynek for his dedicated work to promote objective scientific investigation of UFOs, extraterrestrial intelligent life events, and Close Encounters;

To biologist and astrobiologist Margaret Race, for her curiosity, insights, and scientific expertise as Senior Research Scientist and as Principal Investigator, research and outreach activities on Planetary Protection, SETI Institute, Carl Sagan Center for the Study of Life in the Universe;

To Col. Jesse Marcel, Jr., MD, who continually and courageously fought on behalf of Roswell witnesses to promote historical accuracy so that the truth about Roswell would be known;

and

For Janie, with love.

CONTENTS

Acknowledgments | ix

Introduction: The Aliens are Coming, the Aliens are Coming! | 1

1 Aliens and other Others
Where Did They Come From? | 31

2 Extraterrestrials: Theirs and Ours
Are We the Others We Don't Want Others to Be? | 71

3 ETI: Roswell, Riverine, and Rendlesham Encounters
We Are Not Alone . . . ? | 106

4 Xenophilia and Xenophobia
Good Alien or Evil Alien from Space? | 134

5 ETI Encounters: State, Academia, Church, and Science
Adversarial and Accommodating Attitudes toward UFO Reports | 167

6 ETI in *Avatar* and *District 9*
Aren't We Always the "Good Guys"? | 206

7 Spirit, Science, and Space
Who Are the "True Believers"? | 236

Conclusion: *We Aliens Are Going! We Aliens Are Coming!* | 279

Select Bibliography | 289
Index | 293

ACKNOWLEDGMENTS

ENCOUNTERING ETI RESULTED FROM many years of thinking about terrestrial-extraterrestrial intelligent life Contact. As I came to realize how complex the topic is, I expended far more hours doing research and considering various approaches and themes than I had anticipated. I am grateful for the insights I have received from conversations with friends and acquaintances near and far while I worked on the manuscript, and for the ideas of numerous authors I have cited.

I thank my Wipf and Stock editor, Charlie Collier, for his support, patience, and comments as he endured deadlines missed on both *Cosmic Commons* and *Encountering ETI*. I hope you can sleep more easily now. I am grateful to Jacob Martin, my expert copy editor, for his dedicated, conscientious, and insightful work while reviewing the manuscript.

I appreciate the excellent endorsements I received from exceptional people for *Cosmic Commons* and *Encountering ETI*. In the current academic and social climate, the issues I explore are "far out" in more ways than one; you have been courageous and supportive in endorsing their consideration even, at times, when your perspective was very different from—and even contrary to—my own.

I received strong support and loving encouragement—as well as insights during our conversations and debates—from my daughter, Shanti and my son, Daniel. You helped me clarify my ideas, and explore topics with a wider audience in mind.

I am especially grateful for the love and encouragement of my wife, Jane Morell-Hart, and for her patience as I worked on research and writing long hours in the day and late into the night. I think she thought I was becoming too "spaced out" as I did my research and writing, and thought about its implications. It's done!

INTRODUCTION

The Aliens Are Coming, the Aliens Are Coming!

THE HEADING ABOVE MIGHT startle the reader. I want to assure you immediately that this is not a warning about extraterrestrial beings that I've spotted with a home telescope, or about desperately poor potential immigrants seeking to cross the border from Mexico into Arizona; nor is it an advisory to California Department of Agriculture personnel that some earnest plant lover will attempt to smuggle across the state line an exotic ornamental species that will become invasive and threaten the survival of food crops. Rather, the title implies to some extent all of the above. (I must confess, however, that I have no telescope, am sympathetic to diverse people attempting to escape pervasive poverty and political persecution, and support government warnings against transporting potentially invasive plant species, and state government interdiction of these alien invaders at state borders.)

The title should not be taken to suggest, either, that most or even many astronomers, astrobiologists, or astrophysicists think that Contact with intelligent extraterrestrials is imminent. Most scientists theorize that Earth inhabitants' exploratory ventures into space (with or without personnel and passports aboard) will encounter very primitive, microbial forms of life; they'll be in the early stages of evolution, assuming that planetary conditions allow for evolution to occur at all (a possibility which, given recent astronomical discoveries of planets in distant places in space, is becoming ever more likely).

If you're in academia (as I have been for more than forty years), you probably bought this book online or surreptitiously in your local (or out of town) bookstore. You might be secretly reading it late at night, in a closet, with a flashlight. My sympathies. I recognize, as you do, that the U.S. government is unhappy about those who speak about UFOs publicly and thereby contradict its continuing claims that there's no such thing as UFOs, that there was no debris from the crash of an "alien disk" at Roswell, New Mexico in 1947, and that there's no need for an independent, objective scientific investigation into UFO incidents reported by credible witnesses. I recognize, too, that academic institutions have fallen in line with U.S. government propaganda, and punish faculty professionally if they even mention casually that they think serious scientific research should be undertaken on UFOs (unidentified flying objects) and ETI (extraterrestrial intelligence). Finally, I understand that even family members, friends in your social circle, and professional colleagues will dismiss your interest and ridicule you for even mentioning it. (Imagine your quandary if you've actually seen a UFO whose presence, location, and maneuvers—flying horizontally at extreme speeds and then, without pausing or banking, shooting straight up at the same extreme speed, following a vertical trajectory perpendicular to the horizontal one—don't correspond to any natural phenomena, including meteors, the moon, Venus, and nations' satellites).

Relax a bit. Keep fresh batteries with you, and keep reading. We'll try to make this a pleasant excursion into the cosmos—or, at least into what people like you, your next door neighbor who's a pilot, the clerk at the supermarket, and scientists and radar operators—military and civilian—around the globe have reported. There are thousands of them!

When I've spoken about my interest in the topic at faculty meetings or informal gatherings, people roll their eyes or remain silent, among other reactions. I know the eye rollers might be either genuine skeptics or people who have seen UFOs who are afraid that others might detect this and attack them; the silent ones are people willing to think about UFO narratives but dare not say so—even if they, too, have seen one or more. My experience has been that one or several people in the second group will quietly approach me discreetly and tell me about a UFO experience that they have had or someone they know who is trustworthy and truthful has had. I wonder how many people there are who have similar experiences and are similarly afraid to mention or acknowledge them.

The cover story of the July 2013 issue of *Scientific American* might help your cause when people say there is no extraterrestrial life. You can point out that if that were the case, private industry, scientific research organizations, and even the U.S. government would not spend billions of dollars to search the distant skies. The magazine's cover reads, "To Seek Out New Life: Watching exoplanet skies for signs that something is out there"; the story's title reinforces the cover: "The Dawn of Distant Skies—The galaxy is teeming with planets. Scientists are straining to peer into their atmospheres to seek signs of extraterrestrial life." The article provides examples of scientists around the world participating in the hunt, and of the papers they have presented at professional conferences, and notes how advanced technologies have enabled scientists in just under two decades to locate planetary candidates that might have water and to detect atmospheric biosignatures that indicate the presence of life.

What's an "Alien"?

The term *alien*, as seen in the preceding paragraphs, has a variety of meanings: biological/ecological, ethnic/class/political, and extraterrestrial. *Encountering ETI* explores diverse aspects of distinct issues relevant to all of the above, noting a consistency in meaning of *alien* as "outsider," an individual or species not native to a place; and, a corresponding consistency of complementary impacts of diverse types of actual or potential "aliens" in Earth contexts. For purposes of this book, a working definition of *alien* is as follows:

> *Alien* describes an individual, species, or ethnic group that enters a territory not native to it, which is inhabited already by members of the same or a similar species or group that currently utilizes subsistence natural goods and habitable space of their current place in a biotic niche to which they have adapted; that some or all of the goods and part or all of the territory will be sought or affected by the immigrant species, in competition or collaboration with the natives; and that the natives' space and subsistence goods, and therefore their likelihood of survival or wellbeing, could be adversely impacted or beneficially enhanced by the nonnative biotic immigrants.

In the pages that follow, we'll ponder ecological, economic, ethical, and ecclesial theoretical and actual engagement with aliens, employing a long-term eonic lens. We will view and review the evolutionary past, evolving present, and ponder potential evolutionary futures. We will analyze species' coadaptive and integrated—or conflictive and divisive—relationships with Earth and with each other. We will discuss real, or posit potential, results of alien migration—on Earth, to Earth, and in the far reaches of the universe.

You'll Do All That in One Little Book?!

The task, of course, is enormous; it requires several limitations. Geographically, we'll focus on U.S. settings; there's certainly enough going on here to merit substantial on-planet exploration. We'll look at several issues in summary fashion, in hopes that the reader and others will expand consideration of these issues, and also extrapolate data and insights from what is presented here to reflect on related or even apparently unrelated issues. We'll use the word *Contact*, capitalized as in scientific circles, to signify encounters between terrestrial intelligent beings (TI) and extraterrestrial intelligent beings (ETI), which has often been shortened inaccurately to ET (which means "extraterrestrial," and could refer to exoEarth life that is simple—such as microbes—or complex—such as dolphins or their intelligent equivalents on other worlds).

Natural Goods and Natural "Resources"

An important distinction needs to be made between "natural goods" and "natural resources" and how we relate to each. A *natural good* is something that has a place and serves a purpose in its native setting; it may be altered or moved when this is necessary to provide some benefit. It might remain where it is (not in its entirety, necessarily: think of river water that is drunk immediately by a hiker to satisfy thirst, or is used to fill a canteen to be drunk later; or is partially diverted by an agriculturalist—farmer or rancher—into a canal to provide irrigation for crops or sustenance for livestock, both of which people will consume later), or it might be diverted elsewhere to flow from a home faucet—after it has been purified in a municipal water plant in a nearby community—to wash vegetables produced by the farmer, or when added to dehydrated vegetables to make soup or stew.

A *resource*, by contrast, is something that is regarded as awaiting alteration, extraction, diversion, or other re-use by human ingenuity, labor, and technology. While a "natural good" is respected for what it is, even when altered or removed, a "resource" is regarded as something to be changed at human whim or will: what it will become is valued over what it is now.

Introduction

A "natural good" is understood to have *intrinsic value*, worth in itself, and is worthy of respect; its existence is primary, and complementary to or even prioritized above human needs and wants when it provides for other biota (living beings, members of the biotic community, the community of all life), or has purposes in nature only partially known at present. A "resource" is believed to have *instrumental value*, a worth assigned to it by those who want to use it; its existence is secondary and subordinate to human needs and wants. A being's intrinsic value might come to be displaced in the eyes of others to become instrumental value when, in their view, they need benefits it can provide—whether to satisfy their needs or desires—to sustain their own intrinsic value. A thirsty and hungry grizzly bear or golden eagle on a river in the Pacific Northwest, for example, depends on water to slake its thirst and to be host and habitat for salmon to satisfy its need for food. The salmon, for its part, has intrinsic species and individual value, eats insects on or above the water's surface because they have instrumental nutritional value for the salmon, and has instrumental value for the bear and eagle as food. A similar intrinsic value–instrumental value relationship exists between the salmon and insects that it eats, some of which might eat the salmon's decaying remains after it spawns and dies.

Throughout *Encountering ETI*, I will use "natural goods" rather than "natural resources" to refer to what exists integrally in place, sustaining abiotic geodynamics or providing benefits for resident nonhuman biota, but might be needed and used by human beings. The use of "natural goods" to refer to Earth benefits that humans use can promote respect for Earth and the biotic community, and responsible use and distribution of Earth's geophysical places and their fruits.

Intrinsic Value and Instrumental Value

In ethics, as noted, both biota and abiotic places can be acknowledged to have intrinsic value (value inherent in themselves), or assigned instrumental value (value that benefits in some way the one doing the valuing). Note that in the first case a person who thinks ethically acknowledges an inherent value, but does not decide that the biota or abiota has it; rather, they understand intrinsic value to be something that is internal to and part of the being (which might have been imparted by their common Creator) of the other, not something to be granted by others; in the second case, the one valuing assumes that they have the right to do so, to benefit themselves, their species, or their community.

Environment in these pages refers to places and spaces that are the common ground (and common air and water) where geophysical forces (such as tectonic shifts, climate, and storms) exist and interplay. *Ecology* describes the relationships that exist or should exist in Earth and cosmic environments: among humankind, among all biota, between humans and other biota, between humans and Earth, between other biota and Earth, and between humans and other biota, related together to Earth.

Earth: Home, Hearth, and Habitat

In *Encountering ETI*, as is customary today in contemporary scientific, ecological, and spiritual writings, I capitalize our home planet, Earth. The uppercase E distinguishes Earth from the soil, earth; reinforces its status as a planet—all other planets are capitalized; and promotes respect and care for Earth's environment and for other beings on Earth, and stimulates cooperative and collaborative ecological relationships with abiotic (nonliving) and biotic (living) existents. (In some indigenous cultures Earth, rocks, and other beings that research scientists consider nonliving beings are understood in native thought and spirituality to be living beings.)

Earth is *home* to diverse species. It is, for humans and other biota, our residence, the place in which we live and from which we provide for our life and wellbeing—as individuals, families, and communities. As our home, Earth is our nurturer, too, the place we find needed food, water, and shelter, among other goods. It is the world with which we are familiar and in which we have roots. It has a welcoming familiarity that comforts us or puts us at ease and enables us to feel secure (to the extent possible: it's always a cautious security when you're prey for a lurking or tracking predator) when we return to a particular place from which we have traveled.

Earth is *hearth* to a particular species, humankind. We have a particular sense of place here, an experience that seems to go beyond other species' affinity for an established territory. We can be territorial—witness international boundaries and borders between nations—but we can go beyond that to recognize our kinship not only with others of our kind, but other members of the extensive biotic community, the community of all life.

Earth is *habitat* for all species. It is the planet from which, and place in which, we grow and gather nutrition for our sustenance, attire to protect us against diverse elements, material to construct our residences and places of employment, and medicinal plants (or replicas thereof) whose properties provide for our health.

As our home, hearth, and habitat, Earth is the place where we hope to live intergenerationally as a species, where abundant natural goods will enable us to live harmoniously and well. On Earth, we recognize that we are interrelated, interdependent, and integrated members of the *biotic community*, the community of all life.

Interdimensional Ecological Existence

In *Encountering ETI* interdimensional ecological aspects of existence are elaborated: *materiality* (relationships with Earth, other humans, other biota); *sociality* (relationships between diverse human individuals and distinct human communities, and between humans and other intelligent beings); and *spirituality* (relationship with the Spirit).

In the cosmic context, these diverse types of *relationality* are well expressed in the Hindi salutation *namasté*, which has multiple, intertwined, and integrated meanings: "the Spirit in me greets the Spirit in you" (that is, divine being is present in each of us, permeates all beings and every aspect of existence, and is self-communicating in and to all); the spirit in me greets the spirit in you (our individual materiality shares a common spiritual aspect of and relationship in our being); the Spirit in me greets the spirit in you (the sacred Presence in me embraces your spirit); the Spirit in you greets greets the spirit in me (the sacred Presence in you embraces my spirit). Those familiar with the theology of the early Christian scholar-abbot-mystic Saint Maximus (580–662; highly regarded in both Eastern and Western Christianity) would note the complementarity of his core ideas and *namasté*. Maximus wrote and spoke about the dialogic relationship between *Logos* (the eternal Creator) and *logoi* (all being and beings, which have a common origin in divine creative power).

Science, for its part, expresses—in theories and data about the origins of the existing and inflating universe in a singular event called the "Big Bang" in popular thought—a similar concept of how all that exists is related. The holistic understanding of the interrelationship of biota and abiota is present in the Genesis 2 creation story, which describes an original garden paradise. This theme is captured in the 1960s song "Woodstock," written by

Joni Mitchell and popularized by Crosby, Stills, Nash, and Young: "We are stardust, we are golden / and we've got to get ourselves back to the garden."

In Lakota ("Sioux") and other native cultures, *namasté* is expressed in a complementary greeting, *mitakuye oyasin*: "We are all related." In its extended form, Indian elders pray, "Greetings, all my relations. Greetings to all the two-legged people. Greetings to all the four-legged people. Greetings to all the winged people. Greetings to all the finned people. Greetings to all the rooted people." All of these greetings are voiced with the understanding that all creatures exist in the presence of the Creator Spirit. In a different way, relationality is described, too, in science in quantum physics, and in the social sciences of sociology, anthropology, and psychology.

We living beings are all stardust become material, interrelated, interdependent, and globally and cosmically integrated beings and being.

Stephen Hawking's *Deus ex machina*

The catalyst for me to write *Encountering ETI* and its related book, *Cosmic Commons*, was statements made in Hong Kong (2006) and Cape Canaveral (2007) by eminent British scientist Stephen Hawking. Reflecting on deteriorating ecological conditions on Earth, Hawking declared that human survival required development of a moon base and a Mars colony within decades. Earth, he said, might be destroyed by disasters such as "sudden global warming, nuclear war, [or] a genetically engineered virus," and some humans should be resettled: "I think that getting a portion of the human race permanently off the planet is imperative for our future as a species." It struck me immediately as I read his words: *It's the same people!* Those who would be involved in such human settlements elsewhere would be members of the species that is wreaking havoc on Earth. Why would they do anything different on the moon or, more importantly, on Mars and on other "celestial bodies" (a United Nations term, from the 1966 Outer Space Treaty, that refers to the moon and other places in the universe)?

Most journeys of exploration are funded not for solely scientific purposes, but with a commercial or military intent. The Spanish monarchs funded the 1492 voyage of Christopher Columbus that led to the "discovery" (native peoples were already here) in 1492 of what came to be called the "Americas" not as a scientific journey (unlike, by contrast, the case centuries later when science was an important part of the voyage of HMS *Beagle*, with naturalist Charles Darwin aboard) or an anthropological quest to seek or understand human life in foreign countries. Columbus was funded to seek new, economically profitable oceanic trade routes, to expand Spanish territory, extend the influence of the emerging Spanish empire across the globe, and to acquire through all of this additional wealth for Spain. When explorers such as Columbus encounter territory where they find "resources" useful for them and their patrons to meet domestic needs and wants or for trade and commercial profit, substantial military personnel accompany subsequent voyages to "discovered" places to ensure security for colonial expansion and control. The imperial set of political, economic, military, and even religious forces will strive to secure access to and control over regional planetary goods, and to subjugate "uncivilized" peoples ("uncivilized" in the perspective of expansionists who define "civilization" strictly in terms of their own culture) to the colonizers' imperial needs, aims, and domination.

Exploration and attempts to conquer are rarely (if ever) accomplished while bearing in mind any respect for or accommodation to existing populations, or concern for ecosystem integrity, which leads to the question: How would human settlement on the moon, Mars, or elsewhere in the solar system, galaxy, or vast cosmos differ in intent or practice from prior human practices during Europeans' (and other cultures') Earthly expansion? Would—or could—Hawking's projected settlements be refreshingly different from current human conduct in the new milieus in which they will, literally, *take* place? Human explorers and settlers might well take the places of—replace—existing biota; take the territory of existing intelligent biota; and take and diminish the natural goods of their newly settled place(s) to satisfy their needs and wants via commercial, industrial, and social exploitation (the 2009 film *Avatar* illustrates well the human potential to follow this course of action on other celestial bodies).

Will Hawking's seeming confidence (that on distant worlds human thought and actions will evidence greater consciousness of the potential double effect of technology, its use for good or for ill) be justified in the human future? He advocated technology's use to save our species from its abuse in the past and present. But he did not mention humans' technological development of nuclear weapons, massive strip-mining machines that rip apart Earth's mountains and plains and harm biotic habitats and ground water, utilities' power plants that poison the skies with emissions, and manufacturing plants that pollute water with toxic effluents. Other than on Earth, Hawking seems to believe, humans will use technology responsibly, even though on Earth it is destroying the planet, disrupting the social order, and catalyzing celestial colonization. Ironically, Hawking's litany of reasons for resettlement provides dramatic examples of how technological innovations have had catastrophic consequences. (This is not an attack on or indictment of technology *per se*, but on technology's abuse and misuse. In engineering, computers, household appliances, aviation, and medicine, among other areas of human inventiveness and endeavors, technology has enhanced our lives. I much prefer my word processor to my typewriter of decades past; on occasion, I remind my daughter, son, and students that it is much easier for them to write essays and papers than it was for me.)

Hawking proposes a new kind of *deus ex machina*: today's gods will be technological marvels that will save the species as they ship settlers to contained colonies, bringing salvation to a selected few, the elect members of our species (without designating who among us will be chosen, and by what criteria) from what we have wrought on Earth.

Those Not Left Behind

Hawking's scenario is eerily reminiscent of the *Left Behind* novels authored by evangelicals Tim LaHaye and Jerry B. Jenkins (this is especially ironic, since Hawking is an atheist). They believe that a *deus* sans *machina* will whisk true believers "up" to heaven, away from a soon-to-be-destroyed or drastically altered Earth. A popular (among true believers) bumper sticker stated, when the novels were written (and still seen on some bumpers, despite the failure of that then-latest "end of the world" prophecy), that "When the rapture occurs, this car will be empty," which left one wondering about consequences of empty cars careening down highways and crashing into unrepentant sinners' still-occupied vehicles.

In Hawking and the novels a select few are transported into space to escape from Earth's ecological and social destruction. Who are these few? In the novels, the "saved" are those whose particular religious ideology claims that all they needed was faith in Jesus and additionally, for some, close attention and obedience to the Bible. In Hawking's proposal, no criteria are elaborated. The saved few in his scenario are likely quite distinct from the novels' raptured few: they might, for example, be atheist scientists, or people whose sexual morality would be called into question, for one reason or another, by Christian fundamentalists; their social morality might be equally or more questionable by others: they might be characterized

Introduction

by greed for wealth or a lust for power, to be satisfied by whatever means possible.

Those Left Behind

The end of the world stories by Christian fundamentalists were complemented by a different type of "end of the world."

As the year 2000 approached, in response to "end of the world" predictions and "prophecies," the "true believers" in the dogmas expressed in the LaHaye-Jenkins novels anxiously or ecstatically awaited midnight's aftermath. Simultaneously, reacting to Y2K fears anticipating computer clock failures, computer systems users who were "true believers" in technology convinced similarly thinking individuals, governments, and businesses to expend substantial funds to save their data and operating systems; as midnight approached they huddled around individual or corporate monitors nervously drinking massive quantities of caffeinated liquids while wondering, would the "fixes" work or not?

The Christian and computer predictions and concerns were followed in turn a few years later by a New Age prediction: that a "Maya calendar" foretold the end of the world on December 12, 2012 (12/12/12)—which also fizzled, this time leaving New Age true believers and others relieved. Instead of predicting anew, they focused on an aspect of the prophecy upon which others had focused: a revolutionary change in global human consciousness would be catalyzed, in the thought of some fans of ancient Maya culture, by the arrival and teachings of benevolent extraterrestrial intelligent beings.

In the dramas of ancient Greece, when the hero-protagonist had been cornered in some confined space and no escape seemed possible, the *deus ex machina*, the "god of the machine," was lowered by ropes and pulleys to rescue him. In the play then, the simple machine carried the god who rescued the hero; in real life today, the machine *is* the god: technology will ensure human survival. However, even if the new machine-god carries arbitrarily selected people away to safety, a necessary spiritual and social conversion in human consciousness, conscience, and conduct will not miraculously emerge and accompany Earth's survivors in space. Minds and hearts need to be transformed *before* departure; change will not develop *ad hoc* during extraterrestrial extension. *Avatar* illustrates in parabolic fiction a perspective that contrasts sharply with Hawking's. It presents dramatically the kind of human-caused social disintegration and ecological destruction that will occur on distant celestial bodies if humankind continues with its current mindset and the kind of behavior that expresses it.

Hardin and Hawking: "Lifeboat" and "Lifeship" Ethics

Biologist, human ecologist, and professor at the University of California-Santa Barbara Garrett Hardin became very concerned about the impacts of human overpopulation on an imperiled planet. He sought to diminish and then eliminate the pollution problems and natural goods scarcities that were beginning to develop on Earth. He reasoned that other than by strict planetary birth control practices or, that being unfeasible, strict national immigration policies coupled with birth control requirements (expressed through maximum allowable offspring limits) humankind in the near future would suffer from polluted air, land, and water, lack of life's necessities, and ongoing conflict. He had a bias in favor of the "haves," whom he did not take to task for their consumerism, and blamed the "have nots" for their irresponsible birth control practices and for their poverty. He did not blame wealthy nations, corporations, and individuals for their role in causing Earth's problems and in keeping the poor, poor. (The latter phrase is taken from something Brazilian Archbishop Hélder Câmara said some years ago: "When I fed the poor, they called me a saint. When I asked, 'Why are the poor, poor?' they called me a communist." The upper class and its controlling government did not like socioeconomic structural questions to be raised.)

In the article "The Tragedy of the Commons," Hardin used the idea of an agricultural commons wherein different farmers grazed their livestock to illustrate that while this worked fine with a small farming community, as the number of farmers and their family members grew the commons would be strained beyond capacity to sustain them. Two factors caused this: the increased number of farmers, and the desire of each farmer to graze increasing numbers of cattle to attain additional income to satisfy needs or wants. Hardin rightly dismissed Adam Smith's "invisible hand" economics, declaring that it must be "explicitly exorcized." But then, after narrating the cattle herdsman story to show how this "invisible hand" does not promote community economic stability the way Smith theorized (and which, it should be noted, has been used for centuries as a mantra by ultra-conservative economists and politicians to reject minimum wage and public health laws, and oppose social programs that benefit the poor), he limits his dismissal solely to population issues. He asserts, based on the herdsmen story, an inevitable opposition between freedom and the concept and practice of a "commons," because "as a rational being each individual herdsman seeks to maximize his gain." Hardin states as an absolute, based on the story: "freedom in a commons brings ruin to all."

The foundation of Hardin's blanket assertion is flawed in its universal declaration of an inherent human selfishness, disguised as self-interest. Hardin's "rational herdsman" might learn as an individual or through a community's shared consciousness that the wellbeing of all is a community value, and the implied or active social pressure that results from this view would prevent the individual herdsman from either continuing or even initiating such anti-social action. The herdsman's "freedom" in such a context might then be a conscious or subconscious "responsible freedom," not the irresponsible license assumed by Hardin and his source, William Forster Lloyd (1794–1852). In his responsible freedom as a member of a community whose other responsible members are committed to community and a commons for all, the herdsmen as individuals and as a group would note, if not foresee, potential disastrous consequences of each seeking to maximize individual financial benefit at the expense of their commons and their community relationships and wellbeing. There is not then, contrary to Hardin, an inevitable contradiction between "freedom" and "commons," and an inevitable conflict when rational people consider how best to provide for their livelihood in community. The Basque Country in Spain has a marvelous example of a seventy-year-old cooperative, the Mondragón Movement, which integrates agricultural, industrial, trades, service, and other member cooperatives to benefit each cooperative and all individual members of all cooperatives—and Basque society, as a whole. The people-based, -owned, and -operated Mondragón embodies "responsible freedom," "community," and "commons" concepts, values, and practices. In an individualistic economic ideology such as capitalism that uses Smith's image as its idol and where greed is transformed from vice to virtue, there might indeed be conflict between "freedom" and "commons," to the eventual detriment of both communities and individuals.

In human population issues, the principal area of Hardin's scholarly expertise that is the focus of this and the next chapter, Hardin's assertions about (irresponsible) freedom of offspring choices and its resulting social consequences would be welcomed by China, human population limits organizations, and individuals concerned about adverse social and environmental impacts of geometrically progressing human population growth: he states that governments must intervene to limit individual/family freedom of choice on the number of children they would be permitted to have. In Hardin's words, "To couple the concept of freedom to breed with the belief that everyone born has an equal right to the commons is to lock the world into a tragic course of action." He deplores the United Nations' Universal Declaration of Human Rights statement that "any choice and decision with regard to the size of the family must irrevocably rest with the family itself, and cannot be made by anyone else." Hardin proceeds from his disapproval of that statement to proclaim that "if we love the truth we must openly deny the validity of the Universal Declaration of Human Rights." As with his own declaration that "freedom" and "commons" are incompatible, based on one fictitious story by someone Hardin labels an "amateur economist," Hardin extrapolates from his interpretation of one phrase of the U.N. Declaration to reject the entire document, a document accepted in principle if not so much in practice by the nations of the world; Hardin thereby presents a logically indefensible and questionable leap to make in any statement or argument. He concludes his essay advocating government coercion to enforce population control. People (in reality, the dominant culture and dominating social class, which he affirms and celebrates) must acknowledge "the necessity of abandoning the commons in breeding.... Freedom to breed will bring ruin to all."

In his provocative article "Living on a Lifeboat," Garrett Hardin continues his consideration of the consequences of unrestricted human population growth. His ideas have been denounced by churches and other faith bodies that teach that human sexual intercourse is solely or primarily for procreation, and is not to be used with a primary or sole purpose of expressing love—even between a husband and wife who already have several children. Unrestricted intercourse for procreation would, however, augment current human overpopulation, would strain and eventually break Earth's ability to provide for human (and other biota's) needs for natural goods even for survival, and would exacerbate Earth's environmental crisis through pollution, water loss, and soil depletion; the foregoing factors, working together, would catalyze a scarcity of essential natural goods and would imperil humans' survival. Hardin proposed that humankind—or, rather, those humans who are in a nation the majority of whose members are secure in their financial wellbeing and have sufficient natural goods available within their national borders—needed to develop a "lifeboat ethics" in order that select members of the species could survive; newcomers would be excluded—immigrants and unregulated offspring from either native citizens or specialized and limited categories of "aliens." If the poor at home and abroad continue to increase and multiply, there will be a continuing drain of necessities for all people because the excessive numbers of poor people will make demands on Earth's and humans' goods within their own and others' countries (Hardin's solution: severely limit foreign aid, particularly in the form of food). "Metaphorically," Hardin states, while in actuality expressing socioeconomic realities throughout the planet, "each rich nation amounts to a lifeboat full of comparatively rich people. The poor of the world are in other, much more crowded lifeboats." He continues: "each lifeboat is effectively limited in capacity. The land of every nation has a limited carrying capacity." Hardin blames the poor themselves for their poverty, and wants no part in looking for past and present economic injustices that forced them into poverty and to be at the economic disposal of the rich at home and abroad: "The concepts of blame and punishment are irrelevant." As a confident member of the dominant culture, individually satisfied as a member of an economically well off social class in the richest country in the world, Hardin avoids considering the important relevant question, "Why are the poor, poor?" and its corollary, "How did the rich become rich?" The lifeboat—a First World nation—should be reserved for those on board at this moment in history, no questions asked about how they came to be on board while others are vainly trying to stay afloat in the ocean around them. Human overpopulation, he asserts, is an Earth-endangering problem; those who do not restrict their own contribution to it should not have others' sympathy or receive help to survive.

Biologist Garrett Hardin and mathematical physicist Stephen Hawking are in agreement on overpopulation. Hawking cites it as an issue that imperils Earth today and will destroy Earth in the next two decades. It is part of Hawking's rationale for rescuing some humans from humans' self-destruction and Earth-destruction.

Stephen Hawking's "rescue plan" for a human species confronted by ecological catastrophe catalyzed by past and present human consciousness and conduct is to select some members of humanity (he does not cite criteria for selection, who might be selected according to these criteria, or who would do the selecting) and to transport them as expeditiously as possible to the moon and Mars. His contemporary version of a *deus ex machina* would be a technological savior whose occupants likely would be designated and guided by politicians, the financially well-off who support them, generals, and scientists. His proposal is similar to Hardin's "lifeboat ethics." The "lifeboat," however, has been enlarged to a "life(space)ship." The different sized and differently purposed vessels serve the same philosophical and ideological agenda and voyage purposes and proposals: to carry (some, select) humans to safety, safeguarding sustenance supplies from, and negating security for, those who were not fortunate enough to be selected or to self-select to be protected from planetary perils. "Lifeboat ethics" in one case, and "lifeship ethics" in the other, enable the few who are economically and politically advantaged over others on Earth in some way—by dint of force or supposed fortuitousness, or even a type of Darwinian "natural selection" or of Spencerian social selection—to escape from life-threatening danger. The lifeboat-become-lifeship will enable them to resettle in a more hospitable environment on another planet or similar setting. Historically, a similar selection process occurred when the ocean liner *Titanic* struck an iceberg and began to sink, and when selection was made of which passengers would be allowed in the limited number of lifeboats, each with its recommended and maximum seating capacity. Who was allowed on the lifeboats, and who was prevented from boarding them—what criteria were used? Was anyone from the lower decks, and thereby the lower socioeconomic class, allowed among the privileged?

Hardin's proposals for saving people on Earth and Hawking's proposals for saving people off Earth share in common selection of an elite group to be saved to a better natural environment. LaHaye-Jenkins share their belief that an elite group should be saved, but the similarity stops there: LaHaye-Jenkins state that the group will be selected by God to be "raptured" away from a condemned Earth to dwell with God in heaven, a supernatural exoEarth and exoCosmos place.

In contrast to the preceding proposals and projects, what Earth and the extended and expanding cosmic creation need for the future is not for threatened and traumatized people to be dis-placed to new worlds, but the transformation of human ideals and ideas prior to, during, and at the end of extraterrestrial exploratory voyages, colonization and settlement, and entrepreneurial ventures.

Stephen Hawking's more recent comments regarding extraterrestrials and human space colonization, voiced in his 2010 BBC documentary series *Into the Universe with Stephen Hawking,* expand on his earlier statements about humans' space voyages. Pessimistically, he wonders if an inevitable result of life's evolution into intelligent beings, wherever it occurs in the cosmos, is that evolved life destroys its home world. On a realistic and historical note, Hawking compares impacts of possible twentieth century intrusive and invasive arrivals on Earth by extraterrestrial intelligent beings to impacts of Europeans in the fifteenth century. He notes that Columbus' arrival in what would become the Americas "didn't turn out very well for the Native Americans." Ironically, here, he still doesn't warn that humans, whom he has said are destroying *their* planet, might destroy the other worlds into which he wants to send them to save the human species—a warning that, by contrast, is made very strongly by the film *Avatar*. Neither does Hawking consider the essential (Christian) Doctrine of Discovery that provided the ideological foundation for European exploration, expansion, invasion, colonization, and ongoing imperialist seizure of native peoples' territory and natural goods.

The concept and impacts of Discovery will be noted periodically throughout *Encountering ETI*. The ideology continues and if not eliminated will govern, as in *Avatar*, human explorers' and colonizers' attitudes toward and actions against indigenous species—including intelligent species—on other worlds as it did on what Europeans called the "New World" of the Americas, where native peoples had lived, farmed, and fished for tens of thousands of years.

Hawking's admonition about what aliens might do on Earth should be taken seriously, but also extended to include what humans might do to settle on and acquire the natural goods of other worlds. We should take his words to heart as we explore the events—and their implications—discussed in these pages, and as we reflect seriously on, and think creatively about, what we envision as Earth peoples to be our role and responsibility on all the common ground we now or will call "home"—on Earth or elsewhere in space.

A corrective for Earth's human-caused or human-exacerbated ecological catastrophes and human-caused social problems would be to integrate technological development and moral commitments in an innovative cosmic consciousness; to express compassionate concern for the extended cosmic community; and to concretize commitment to cosmic care. Prior to departure for the stars, people need to integrate technology and socio-ecological (eco-justice) ethics. They should be concerned, too, with the social, cultural, and physical wellbeing of Earth's biotic community (the community of all life) in its diverse forms—and then act to avert harm to extraterrestrial locales and life when in space.

A dialogic relationship could be established, in that case, between present Earth and projected future planetary settlements. Reflection on and repentance for what humans have done to Earth and, particularly in the case of European colonization of the Americas and Africa, to indigenous populations and to ecosystems, might lead to an ecological (if not spiritual) conversion—native populations on other worlds will not, in the future, be subjugated and oppressed as indigenous populations have been on Earth; natural ecologies on other worlds will be conserved, not contaminated. Humankind, as it considers the extreme ecological destruction it has wrought on Earth, can do better on other worlds—an implied hope of Stephen Hawking.

A "Far-Out" Topic for Our Time?

Today we live on a planet plagued with poverty, pollution, and political strife. In the United States, the economic gap between rich and poor, and the income gap between corporate owners and managers and working people, are at their widest in history. Factories spew toxic chemical effluents into our waters, power plants send toxic emissions into our skies, and agribusiness corporations spray harmful chemical "-ides" (pesticides, which include herbicides and insecticides) and artificial fertilizer onto crops that provide our vegetables, and sift harmful additives, including animal parts, into the feed that is given to cattle, pigs, chickens, and other sources of meat. An undeclared class war promotes poverty and racism (including eco-racism) somewhat more subtly than in past eras. In such a social and ecological setting, why should we consider—how could we possibly take the time to consider—something as apparently "esoteric" or "otherworldly" as Contact between terrestrial and extraterrestrial intelligent life?

Thinking about space can stimulate us to consider more seriously what is happening on Earth, and how we might redress and rectify humans' most egregious violations of human, biotic, and planetary rights and wellbeing. As we think about what we would do differently on other celestial bodies and to biota inhabiting them so that we would not replicate in the heavens what we've done on and to Earth, we might have one of those "Aha!" moments when we wonder why we don't just start changing our consciousness and conduct here and now. Similarly to how, for some

Christians, consideration of afterlife possibilities might prompt them to live more morally responsible lives in the here-and-now (whether they act from fear of punishment or from love of God and neighbor), so, too, pondering exoplanetary possibilities might prompt us all to do better at the present moment.

We are part of an amazing and awesome universe. The Hubble, Kepler, and other NASA telescopes have taken photographs of parts of the cosmos whose complexity and grandeur were inconceivable not only to our ancestors, whose religious stories and myths could not have conceived such a wondrous universe ("the world" was, for them, a self-contained reality, all of which was visible from Earth with the naked eye), but to ourselves, even as recently as the mid-twentieth century. If, as some suggest, humans are the only intelligent life in the entire universe, then we have an even greater responsibility to live life related well to each other in our human family, to the biotic community as a whole, and to our Earth home. If we are not the only intelligent life in the universe, we should consider well how we are to relate consocially and collaboratively, constructively and congenially, with other intelligent beings whom we encounter.

When we link the preceding with the challenging and deteriorating ecological and social conditions of our Earth home, we realize that by considering terrestrial-extraterrestrial interaction we might feel impelled to better Earth and our human communities. We see, too, that reflecting on this is not as "far out" from the demands of our time as it might have seemed to be once upon a time. We want a better homeland; we want peace; we want people to have at least a minimum of what they need nutritionally to survive, and financially to have security. ETI considerations could be, for some, the impetus that pushes them to work harder to make all of these hopes and aspirations a reality, at least for their descendants if not for themselves.

The Descent and Ascent of Humans

While in the Genesis 1 creation story God creates humans last of all biota and abiotic being, and "male and female God created them," Genesis 2, by contrast, relates how God created the male human first, followed by all other animals, and then the female human. There are then, two contradictions presented in the first two chapters of the first book of the Bible. In his theory of evolution, Charles Darwin stated that humankind evolved from previously present biota who themselves were the result of a diversifying and complexifying evolutionary process: humans are descendants of primates. Humans are also ascendant: they have evolved as an intelligent species, and the gap between their intellectual ability and that of their ape ancestors and contemporaries has continued to widen.

In the vastness of space and over its eons of cosmic time, life on other worlds, too, might have evolved to be intelligent life. Extraterrestrial intelligent life (ETI) might be billions of years older than terrestrial intelligent life (TI)—and considerably more advanced biologically, intellectually, socially, and spiritually. Humankind should be aware of and prepared for such a possibility, not only if Contact occurs on Earth but if it happens on other worlds in near or distant places in the cosmos.

Original Sin and Original Sinners

Genesis 2 presents a mythical garden of Eden populated by two humans who live among the other biota with a relational responsibility regarding them (caring for the garden). "Adam" and "Eve" as they have come to be named in translation from the original Genesis Hebrew, live in a bucolic world. They are almost "angelic" in their goodness and in the biblical writer's regard for them. They commit the first and therefore original sin—they do not fulfill their God-given responsibilities in the garden; they do violate God's commandments for their conduct in the garden—and are sent forth into a less hospitable world. Contemporary evolutionary data and theory, by contrast, evidences and suggests that humans are not "fallen angels," as might be metaphorically extrapolated from the Genesis story, but rather "risen apes," in the sense of being primates who evolved to become a more complex and intelligent species. In traditional Christian doctrine, itself evolved from some of the earliest dogmas formulated in the first several Christian centuries, this original sin required that God become enfleshed in human form to save humankind from its original and subsequent sins. In early Christian centuries, theological speculation about God's incarnation in Jesus suggested several ways that God's intent to "save" humans might be fulfilled, such as solely by being born as a human being, and thereby sanctifying humanity as a whole; by making correct moral choices—good rather than evil—whenever the possibilities arose: the very first choice along these lines would atone for Adam's and Eve's choice of evil, and therefore "save" all humankind; by being, effectively, a human sacrifice that would, as in temple animal sacrifices, atone for human sins; or by leading an exemplary life that would stimulate others to do likewise, and then departing physically from Earth. The dominant perspective came to be the sacrificial one, now the subject of increasing debate among Christian scholars, clergy, and laity.

As currently constituted, some biblical and Christian belief implies either *anthropocentrically* (a human-centered way of thinking) that the entire cosmos was affected and infected by the sinful acts of two original humans on one planet in one of its galaxies, or *anthropomorphically* (a process of projecting human characteristics onto another living being, in this case believing that human conduct is symptomatic of all intelligent life conduct) that there was a fall everywhere in the cosmos where intelligent life evolved or was specifically created, and that consequentially (and again anthropocentrically) the Creator God chose to "redeem" or "save" the entire universe of intelligent life by being born in human form on one small planet in the vast creation.

A religious or theological interpretation of available scientific data, in terms of the biblical creation narratives, would be that humans are "rising apes" rather than "fallen angels": in contrast to the story of emergence in an Earth paradise, a "garden of Eden," and falling from grace, humans have evolved from and are descendants of apes, and are an ascendant species. We are "stardust," then, and we "have to get back to the garden": not to a mythical paradise in which our ancestors lived long ago, but to the ideal that the story indicates and toward which it invites us. As we heed the invitation, we will over time build this place continually as we evolve culturally, biologically, and materially-spiritually. Someday our descendants might experience and enjoy it as their home, hearth, and habitat.

Terrestrial and Extraterrestrial

Ordinarily, the pairing of the words *terrestrial* and *extraterrestrial* has come to mean, in our time, a distinction between who or what is from Earth (*terra*) and who or what is from beyond or external to (*extra*) Earth. However, an analysis of the words and the world reveals that if we regard our part of Earth, our shared territory, as "terrestrial," then what is outside of our part of *terra* is "*extra*terrestrial." So, for example, whether our territory be an isolated island or a country contiguous with other countries, we who reside there are the terrestrials and those who cross the boundaries that we have established as our territorial parameters are the extraterrestrials.

In this sense, then, native plants are native species and terrestrial; nonnative plants encroaching on their territory are nonnative species and *extraterrestrial*; they are *alien* to the natives' territory, and become *invasive* species when they begin to displace the natives and take over their habitat and the natural goods (such as soil and water) that they need and are using for their sustenance, survival, and wellbeing. Similarly with human populations: dwelling in our territory, we are native and *terrestrial*; incoming nonnative peoples are *extraterrestrial*, and *alien* to our territory; they become *invasive* people when they begin to displace us or even when they just seek to displace us, to acquire our home and habitat and the natural goods we need for our sustenance, survival, and security. Aliens, if they and native biota do not adapt to each other, will disrupt lives (and, for humans, livelihoods) and be life-threatening for native biota, and ecologically and environmentally disastrous for native places.

When humanity departs from *terra firma* (Earth's "firm ground") into space, to a certain extent voyagers immediately become extraterrestrial: they are *extra terra*, beyond Earth. If humanity were to construct a space vehicle designed and destined to travel among the stars through generations of human occupants, eventually its original inhabitants, and then their descendants, would become accustomed and accommodated to their voyaging "planet." If they were to return to Earth after generations, or, more so, after eons, they would certainly experience being extraterrestrial: they might have some ancient pictorial record of their original home planet, but it would have altered—or been altered—to such an extent that what was familiar to their initially journeying ancestors might scarcely or no longer exist. To the Earth inhabitants of that distant time, these arriving cosmonauts and unanticipated intruders would truly be, even though distant cousins, "extraterrestrials." Similarly, when these space voyagers approached or encroached upon other planets, and if these planets' intelligent inhabitants are acknowledged to be, in whatever language they speak, the "terrestrials" of their home place, there, too, the once-Earthlings would be regarded as "extraterrestrials" and "aliens," and viewed with at least initial suspicion, no matter how honorable their intentions.

All this being said, *Encountering ETI* will explore the meanings of *extraterrestrial*, *alien*, and *invasive* relative to life and living on Earth. Life that enters native inhabitants' territory, after existing "extra" territorially, outside—whether that life be flora or fauna, microbial or mammal—will be, to those native, terrestrial inhabitants—whether flora or fauna, microbial or mammal—extraterrestrial immigrants.

Encountering ETI will use an *eonic* view to explore the ecological, economic, ethical, and ecclesial implications of terrestrial-extraterrestrial intelligent life Contact. An eon is an extensively long period of time; where it denotes a specific number of years (in geology, for example) it means one billion years. Whether we consider the past histories of Earth and the cosmos, or the immense period of time between the singular exploding point when the universe began and when Earth was formed, there have been multiple eons of one billion years each; we anticipate additional eons in the future of the cosmos.

Coercive Conduct and Academic Apprehension

Discussion of whether or not intelligent extraterrestrials' existence, let alone of Earth experiences of Contact, is risky for faculty to explore who are theology schools and religion department scholars or university science and social science researchers. Some scientists, humanist academics, and members of the general public smile in amusement when serious discussions take place regarding possible Contact between terrestrial intelligent life and extraterrestrial intelligent life, in past, present, or future times. Others react strongly against the suggestion that such life exists, greeting it with the same sort of hostility or derision that they express when peoples of faith traditions affirm their belief in a transcendent Spirit. In their mind, neither divine Being nor extraterrestrial beings exist—after all, "Where's the material proof?" Others recoil from considering the possibility that intelligent life evolved or was created elsewhere: humans are supposed to be atop the created order; some would state, in fact, that all creation exists to benefit "man" (yes, they would use the masculine noun, and they would do so, too, when referring to a divine Being).

In academic circles in the United States, some faculty claim that no intelligent person can believe in God, or that in order to do so they must "leave their brains at the church house door" when they enter to worship; and, similarly, that truly intelligent people do not believe that there are indications that extraterrestrial intelligent beings exist—they vehemently object to suggestions that Contact has occurred in places like Roswell, New Mexico, or the Hudson River Valley, New York (let alone in other contexts, where less documentation exists)—when, obviously, this cannot have occurred. Once again they will say, "Where's the concrete evidence?" There's a certain security in declaring that materiality trumps both spirituality and extraterrestriality: only the visible or scientifically and technologically detected material world is real and matters. It can be understood through those of its aspects that can be quantified, qualified, or falsified. There is, too, in some higher education institutions and scholarly circles, the fear of losing research funding if such topics are broached, or of not getting tenure or an expected promotion or salary increase. Such attitudes and fears have been assisted by ongoing government suppression of purported ETI evidence, ridicule of those who want to study it (including using the theories and tools of science), and rejection of claims about ETI experiences; this has helped to legitimate and enforce academic coercion and academics' censorship, even their self-censorship.

Encountering ETI questions the doctrine of a universal or cosmic "fall," and consequently questions aspects of some current understandings of the meaning of the life and ministry of Jesus. In my view, we humans are part of an evolutionary process that is an expression of divine creativity gradually and freely unfolding in the cosmos. Consequently, I think that God did not require a human-divine sacrifice. I think, too, that humans are called by the Spirit to strive to replicate, in their own lives and to the extent possible, the ideas and ideals presented by Jesus (and other messengers of God in all religions).

I hope that for believers or agnostics or anyone on a spiritual or humanist quest my personal religious-theological-spiritual (reader's choice) reflection and discussion on these topics (explored more extensively in *Cosmic Commons*) catalyze an enhanced understanding of the Spirit's words, works, and relationship to creation. This would be true, too, in the Spirit's engagement with intelligent extraterrestrials. Their existence would indicate that the Creator can converse with a variety of evolved beings who have an active and often reflective consciousness, and ETI would appreciate humans' openness to learning about their experiences and spiritual development relative to what I have expressed. For Christians in particular, ideas expressed here can provide an enhanced yet focused view of Jesus, and a greater appreciation for the Creator Spirit's universal (literally!) love for all being. The Spirit's communication with—and even incarnation or inspiring with, in some way—beings other-than-human would indicate that expressions of divine love are not limited to engagement with a single

Introduction

species on a single celestial orb, but expansively embrace, and evince solicitude for, life throughout the universe. An essential doctrine of traditional Christianity that God became part of God's creation on Earth need not be changed at all; what should change are some religions' perspectives and doctrines limiting that possibility solely to Earth and to only one person's unique relationship with divinity.

The creative reflection proposed for evaluating parts of the Bible and traditional Christian beliefs in light of possible Contact will, no doubt, occasion a backlash from some believers. However, it is intended to accommodate, if not reconcile, theories and findings of contemporary science and society. This is not new: Christian churches over the centuries have sought to incorporate the ideas from Copernicus and Galileo that, despite once-believed biblical verses to the contrary, the sun and other stars do not circle Earth. In the twentieth century, too, Christians have tried to creatively reconcile their religious dogmas with theories of evolution formulated by Charles Darwin and his successors, including biblical stories and Christian doctrine that literally and literarily express beliefs that are contrary to scientific facts.

Despite concerns or apparent constraints, your intrepid author has decided to make a "leap of faith" and consider diverse implications of Contact. This will be done particularly in order to reinforce the efforts of communities of faith, scientists, environmental organizations, and the general public to, quite literally, "save the world," the Earth on which we live, and to "save the worlds," the other celestial settings to which humankind will journey (and perhaps devastate in a manner similar to what human consciousness and conduct have done on and to Earth). This is, certainly, a different kind of "salvation" than that taught by traditional Christian doctrine, expressed at its worst by LaHaye-Jenkins, and not a limitation of religious beliefs.

Very helpful words for this discussion are provided by Harvard biologist (and secular humanist) Edward O. Wilson. In his book *The Creation: An Appeal to Save Life on Earth*, he suggests that secular scientists and religious believers "meet on the near side of metaphysics," and states further that science and religion, working together, are the best hope for saving the planet. This thought and proposal can be extended: science and religion should collaborate to save this world and also, preemptively and then on-site, save future worlds that will experience the human presence.

A Twenty-First-Century ETI "Thought Experiment": Exploring Near and Far Frontiers

Considerations of the possible existence of and Contact with extraterrestrial intelligent life (ETI) will be undertaken in *Encountering ETI* as a "thought experiment": a discussion of Contact "as if" it has already occurred or will occur in the near future. Deliberations along these lines will complement (and might to some extent overlap) what I wrote previously in *Cosmic Commons: Spirit, Science, and Space*, published in 2013. Through such a thought experiment, we are stimulated to take seriously the impacts and implications of Contact—on Earth or in the heavens, the places outside of Earth in the vast cosmos.

The chapters that follow will not assume that Contact between intelligent beings from different parts of the cosmos has already taken place. No "pro" or "con" position will be taken when presenting data or narratives related to asserted Contact events. However, in order to focus our consideration of the implications and impacts of Contact in the present moment, in order to help humankind adapt later (physically, psychologically, and psychic-spiritually) as a species if universally undeniable proof is presented, and in order to prompt us to care for our Earth home, each other, and our biotic kin, we will "assume for argument's sake" that the stories are history, that is, factually accurate. Then the narratives will be able more fully to stimulate us to imagine how we would react, on Earth or among the stars, if Contact were definitely made. But, while I will not seek to prove definitively that Contact has occurred, I will use the "as if" to catalyze not just speculative considerations but also concrete changes in our current limited consciousness, and consequent changes in our current and future conduct—on Earth and among distant stars.

So, brace yourself. We'll objectively analyze possible ETI phenomena in *Encountering ETI*. This intended objectivity extends to considering seriously and presenting positively evidence of various kinds of "close encounters" expressed in testimony about such events by credible witnesses (as in a court of law, a "credible witness" included here is someone who is "believable" because of their honesty, clarity, direct and coherent narrative, confidence in relating what they have seen, and education—formal and informal—and life experience background—in general and as related to a particular incident).

As a corollary to the preceding, as people ponder how they could be a threat to extraterrestrial life and worlds if they were to export current Earth

Introduction

ecological and social crises and community conflicts during extraterrestrial colonization, they should consider in their current historical and terrestrial context what socioecological (social and ecological) vision they have as an ideal for colonies in extraterrestrial contexts. Dissatisfied with what humankind has been doing on Earth, they should hope to do better elsewhere in the cosmos *and* on Earth (surmounting Stephen Hawking's pessimism about humanity's survival chances on Earth). Such thinking might result in the establishment of a dialogic relationship between present and future. As people reflect on current harmful human impacts on Earth, consider alternative ethical principles and conduct and an altered consciousness that might prevent similar harm in space, and as they consequently envision a better world on an extraterrestrial site, they might consider: "Why doesn't Earth fit our ideal?" They might, then, envision a renewed Earth, and imagine what kind of interactive relationship might be established between future and present contexts such that the envisioned extraterrestrial ideal and the envisioned Earth ideal inform action on present Earth, and prompt humanity to alter human consciousness and conduct on Earth to *realize*—make real—the envisioned Earth. An ongoing and mutually enhancing dialogue could result: in time, between present and future; in place and space, between peoples on Earth and people or other intelligent beings elsewhere in the cosmos.

Considering Contact, Pondering Possibilities

In the following pages, as we explore together through our eonic lens the implications of terrestrial-extraterrestrial intelligent life Contact, we will envision how we should act in such a situation, explore how our envisioning of the possible future should expand our cosmic consciousness and conduct, and reflect on how we might be prompted in the present era to evaluate our current conduct on our Earth home, and change it conscientiously and creatively. Our "as if" thought experiment will be fruitful, then, in the present present and coming presents—whether or not Contact has occurred or will occur.

I acknowledge at the outset that I am a human rights, biota rights, and Earth rights advocate and activist. I have a particular focus on human communities living in beneficial ecological relationships with other biota in diverse and integrated Earth contexts: a socioecological perspective. I write *from* context, *for* context—and contexts. In *Encountering ETI* and its predecessor, *Cosmic Commons*, I have expressed my increasing interest in extending my thinking beyond Earth to exoEarth contexts. My research and reflections have indicated to me that human anthropocentric consciousness and misconduct on Earth will become their cosmic misconduct or that, alternatively, a socially (and spiritually) transformed humanity's consciousness and conduct will promote both Earth and cosmos wellbeing. We will tend to do in the heavens what we have done and are doing on Earth. I continue to hope for and work toward human transformation.

Chapter 1, "Aliens and other Others," discusses the origins of diverse types of aliens, including places, plants, people, pets, and planetary voyagers, and their actual or potential impacts on native species populations and on themselves as they migrate. Chapter 2, "Extraterrestrials: Theirs and Ours," shows how humans are or become "extraterrestrials" on Earth and in the cosmos, and how all space travelers become especially "extraterrestrial" when away from their home planet. Chapter 3, "ETI: Roswell, Riverine, and Rendlesham Encounters," will explore events in the Roswell-Corona area of New Mexico, the Hudson River Valley area of New York, and the Rendlesham Forest area in England using the "as if" thought experiment described earlier to promote serious, in-depth evaluation of credible witnesses' reports, and of potential impacts of Contact on-site and beyond. Chapter 4, "Xenophilia and Xenophobia," will analyze people's contrasting preconceptions of extraterrestrials, as present in the public mind and in science fiction writings and film: *benevolent* beings who will shower us with technological knowledge and medical wonders, or *malevolent* misanthropes (to stretch the term somewhat) who will try to conquer humankind in order to acquire our planet with its natural goods, and annihilate or enslave humanity; we would tend to "love our neighbor"—practice xenophilia—in the first instance, and fear our neighbor—practice xenophobia—in the second. Chapter 5, "ETI Encounters: State, Academia, Church, and Science," discusses how humans' social institutions have reacted to reports of UFO sightings or ETI Contact, including the U.S. government's denial that UFOs exist, religions' resistance to UFO stories (because such narratives would call into question specific religious dogmas), academic institutions that fall in line with government assertions, and scientist faculty and researchers who reject *a priori* such narratives, requiring physical evidence of their presence. Chapter 6, "ETI: *Avatar* and *District 9*," analyzes the meaning of "alien" in exoEarth milieus and discusses human and other aliens' possible conduct when they land on worlds other than their own and abuse resident intelligent beings, and human conduct when aliens land on Earth and humans abuse them; contrasting principles and practices for human conduct which have been detailed in international documents are presented for consideration. Chapter 7, "Spirit, Science, and Space," will explore the relationship between religion and science as expressed particularly in the lives and writings of creative scholars who walk comfortably in both fields; some contemporary thinkers seek to bridge Spirit and science in space, using ETI reports to provide data to develop a common intercelestial community consciousness. In the Conclusion, "We Aliens Are Going! We Aliens Are Coming!" believe it or not, all of the preceding will be nicely integrated and wrapped up . . . or will it?

Immediately before I began to write *Encountering ETI*, I finished writing the now-published *Cosmic Commons: Spirit, Science, and Space*. *CC* is a more in-depth study of issues in the current book but, at the same time, an extensive exploration of other issues. As it looks at the implications and impacts of Contact between humankind and extraterrestrials, it examines the social and ecological links between what we're doing on Earth now, and what we might do when roaming the stars. It approaches cosmic exploration and possible impacts and implications of Contact between Earth humans and cosmic extraterrestrials from four angles: *ecology* (*inter*species and *intra*species relationships among biota, and between biota and their planetary environment, on Earth and on celestial bodies); *economics* (human utilization of Earth's and celestial bodies' natural goods, as is or as altered; distribution of natural goods; and intergenerational use and sharing of natural goods); *ethics* (humans' right consciousness and right conduct in their relationships among themselves, between themselves and the biotic community on Earth and in the heavens, and between themselves and their Earth home); and *ecclesia* (Christian churches' and other religions' perspectives on interaction with other-than-human intelligent species, other biota in an evolutionary state, and Earth and cosmos as a divine creation-in-process). *Cosmic Commons* is a scholarly book with appropriate footnotes, citations of relevant thinkers in the field and of page references for quotes used, a bibliography, etc. *Encountering ETI* is more accessible for a wider audience: people who are interested in approaches to the UFO/ETI topic, and not concerned about academic citations, only essential content. In their interest, to keep them interested, I have dispensed with academic customs to make the book read more easily and flow more smoothly.

Terrestrial intelligent beings (TI) on Earth who say "the aliens are coming!" refer, in cosmic terms, to the arrival of extraterrestrial intelligent beings (ETIs) on Earth. Terrestrial intelligent beings on other worlds who express the same excitement will refer, in future times, to human extraterrestrial intelligent beings arriving on their home world. We become, in space, extraterrestrials—those who have left their origin world and are not native to other worlds. If we and they have become transformed to a social and spatial consciousness that is socioecologically responsible, recognizes in the "Other" something of ourselves, and is open to what the Other has to offer, then all of us, as TI or ETI in different planetary settings, will have a foundation for a fruitful relationship.

As we go forth on our exploratory interstellar journeys, conscious of our extended analysis of and engagement with the biota and abiotic setting of our Earth home, we should bear in mind our ecological, economic, and ethical responsibilities. Our shared hope as a human species should be that in so doing we will be better prepared both to respond responsibly to whatever biota we encounter on other worlds and also, on our home world, to relate responsibly to other Earth species (all of whom are our genetic kin in the biotic community), and to our Earth home and habitat, our shared common ground.

1

ALIENS AND OTHER OTHERS

Where Did They Come From?

THROUGHOUT EARTH'S BIOTIC HISTORY, residents original to a place have had their life disrupted by the arrival of newcomers. Whether the original inhabitants had evolved in that place to become the first form of intelligent life there, or had evolved elsewhere—including to become intelligent beings—and migrated to their current place, biota's survival depended on adapting to their new context with its geophysical characteristics and existing biotic inhabitants. Even if the original residents did not seek to retain their place only for themselves, or the newcomers did not want to appropriate it solely for themselves, they both needed abiotic and biotic natural goods for survival.

In such times and places of interspecies encounter, if there was a sufficiency of natural goods for both natives and newcomers each would co-adapt to the other. If sufficient territory and goods were not available, residents would know or sense that it was necessary for them to fend off the newcomers to prevent their initial or ongoing intrusion; if the threatening intrusion occurred anyway, the original inhabitants would defend themselves in order not to be deprived of natural goods they needed for sustenance and survival. For their part, the newcomers upon meeting resistance would engage in an aggressive offense against residents in order to attain needed territory and natural goods, and to be able to adapt to both available land characteristics and nonthreatening biota or, conversely adapt them to the newcomers to meet their species needs—alter them in some way—in order to survive as a species. Conflictive competition or co-adaptive cooperation in the newly common habitat, in terms of accommodation to mutually acceptable species-specific living spaces, food and water allocation, and shared use to provide for the wellbeing of both, might well determine the ultimate survival of the original inhabitants, of the immigrants, or of both. Either or both species might not have the capability to migrate to a different place, to alter dietary or den requirements, or to prevail in a life-and-death struggle for survival, should accommodation not be achievable.

As species extended their territory or migrated to different territory, they came to alien places: they were not familiar with the discovered geophysical terrain, climate fluctuations, and every type of resident biota encountered—some of which might prove to be competitors, predators, or prey. The inhabitants were aliens to the newcomers, just as the newcomers were aliens to the inhabitants. The newcomers, however, were aliens in an alien (and, in their perspective, aliens') land. The natives were no longer aliens in territory that they had come to regard as their home, a place that was no longer alien land to them: at some prior moment they adapted to it and adapted it to themselves (for example, converting caves to homes, or nesting in or felling trees to provide living accommodations; or, consuming as food a variety of roots or fruits from plants that previously had been unknown—alien—to them).

Alien Species, Invasive Species

In contextual encounters between biota, the resident species might be simple or complex, as might be the immigrants seeking entry into or acquisition of their space. An obvious and easy example of this is revealed in a common news story, particularly and more frequently reported in the past century in the United States: human expansion of residential, commercial, and industrial structures into more rural or wilderness areas, whether forests, mountains, or rivers has occasioned not-so-friendly human encounters with bears, mountain lions, eagles, and salmon, among other biota already in place. Recreational runners and household pets have been attacked by cougars and coyotes, respectively; hikers have been mauled and even killed by black bears and grizzly bears; home, office, and shopping center picture windows and patio doors have been shattered by deer leaping through them; carpenter ants and termites have found new nutrition to satisfy their focused gourmet appetites.

The "other," then, whether seen as an obstructive occupant of needed territory with its natural goods, or as a belligerent intruder into habitat that currently serves nicely to meet the needs of those who have been historically present (even if "historically" means a matter of months, not the course of centuries), is initially an *alien* species, that is, foreign to another encountered species in this place at this moment in time. In the view of new entrants on the scene, the earlier inhabitants are the foreigners; they are a form of life unexpectedly encountered and engaged that might impede or prevent the migrating species, accustomed to satisfying its expansion needs without opposition, from proceeding in their customary manner. Those living in an established, delineated territorial space (even if indicated only loosely: for humans, with artifacts of some sort placed as boundary markers, or by customary use of a territory affirmed by natives and accepted by others; for other species, scent or another indicator marking a place), might regard the incoming "other" as both an alien species and an *invasive* species. The immigrant newcomers are viewed as invasive if their needs and wants conflict with those of the resident species, and adaptation of each to the other is infeasible (too many nutrition or shelter needs would have to be altered) or impossible (there are no available alternative nutrition or shelter goods, and no possibility of migration by either species because of impediments such as inherent species immobility due to factors that might include species' lacking wings or fins, an altered climate in this place and elsewhere, or imposing topographical constraints).

The long-term prognosis for consequences of resident-immigrant encounters might be either extinction or adaptive integration. Consequently, intentionally or instinctually, species will do what seems best in their efforts at self-preservation or their desire for self-enhancement. They will employ such strategies as are presumed necessary in order for them to exist or to have an enhanced existence. Less complex species could not thoughtfully consider alternatives; more complex species might be able to reflect on the extent of their apparent need or desire for this particular place, or these particular natural goods, etc., and act accordingly. This would result in an offensive "fight *or* flight" by aggressive intruders, to be met in turn by a defensive "fight *not* flight" by current inhabitants.

If we think of "alien" in its multiple usages, essentially it is what a biotic being regards and designates as an abiotic "other" place into which it is entering or considering entering; a biotic "other" not previously native to this particular place or not encountered previously during a migration; or a transplanted immigrant who has been accepted as a resident of a place, who might continue always to be an alien, or eventually be regarded as "one of us" as long-term residence continues or as generations pass (assuming that no internal conflicts emerge, or external pressures are exerted, resulting in the "other" being relegated again to an "alien" status). The designation of an arriving or resident biota as an apparently or actually threatening "other" is not necessarily verbal: there are no cell phones or even audible vocalizing of another sort by which individuals in some species can call others of their species to warn them about, and call them to respond expeditiously and aggressively to, if necessary, the incoming "other." There are, of course, many land animals, birds, sea creatures, and insects that do vocalize alarm about known predators or unknown biota entering their accustomed territory.

The types of living "aliens" considered here will be plants, insects, animals, people as a particular species of animals, and exoEarth biota. The preceding includes biota that are complex or simple, intelligent or primarily instinctual and nonreflective. (The most comprehensive regional book for the U.S. in terms of nonhuman aliens is George W. Cox's already classic *Alien Species in North America and Hawaii: Impacts on Natural Ecosystems* [1999], from which some of the following data is drawn.)

Biotas' Ecological Relationships

As noted above "alien" plants might be, for a migrating species, unfamiliar vegetation growing in and around a place which they hope to colonize. The vegetation in such encounters is native and likely evolved there. Perhaps it had been carried there in some form (such as seeds on hoofs, in droppings, or on the wind) by animals, birds, or the elements. As migrants become residents, they become native to a place, and the once-alien vegetation is now familiar and some of it is likely being used for food or medicines (as it is, raw; or, after alteration and preparation by heating to varying degrees).

In *On the Origin of Species* (1859), biologist (and former divinity student) Charles Darwin describes the intricate yet straightforward co-adaptive and coevolutionary relationship between house cats and red clover. (I'll bet that caught your attention! There are, of course, other species involved.) Darwin had noticed that there was a more extensive population of red clover near rural villages, and wondered why this was the case: the clover seemed to grow better with human neighbors than out in the country isolated from humans. In his own words:

> From experiments which I have tried, I have found that the visits of bees, if not indispensable, are at least highly beneficial to the fertilisation of our clovers; but humble-bees alone visit the common red clover (Trifolium pretense), as other bees cannot reach the nectar. Hence I have very little doubt, that if the whole genus of humble-bees became extinct or very rare in England, the heartsease and red clover would become very rare, or wholly disappear. The number of humble-bees in any district depends in a great degree on the number of field-mice, which destroy their combs and nests; and Mr. H. Newman, who has long attended to the habits of humble-bees, believes that "more than two-thirds of them are thus destroyed all over England." Now the number of mice is largely dependent, as every one knows, on the number of cats; and Mr. Newman says, "Near villages and small towns I have found the nests of humble-bees more numerous than elsewhere, which I attribute to the number of cats that destroy the mice." Hence it is quite credible that the presence of a feline animal in large numbers in a district might determine, through the intervention first of mice and then of bees, the frequency of certain flowers in that district!

In this ecological relationship described by Darwin, humans who were once aliens to a place, as were their pet cats, and the clover, mice, and humble bees native to that place, have co-adapted to the changed circumstances caused by immigrant humans' arrival with their feline "livestock" (of course, they might have brought a cow for milk, chickens for eggs, dogs as sentinels, horses for transport, etc.—all of which would originally be aliens too). Since the diverse species have learned to coexist, as described by Darwin, their relationship has become mutually beneficial.

Darwin's narrative also illustrates how species that are seemingly unrelated, or distantly related but more complex (e.g., humans), might regard the less complex (e.g., bees) as unimportant, without intrinsic value (value inherent in themselves) or even instrumental value (value for and to be used by others as food, to provide shelter, to be used for clothing, etc.), even though, in reality, they are all intimately intertwined, integrated, interdependent, and interrelated. In recent decades, people have been seeing even more clearly and extensively that all beings—not just humankind—have intrinsic value as they are; further, they are often beneficial to meet other biota's or an ecosystems' needs, and even provide regional or global requirements for the general integrity of the planet (think about a forest "as is," whose trees provide shelter for bird nests, shade for hot people, dogs, and other mammals, wood for home construction and, ever more importantly, consumption of harmful carbon dioxide and release of beneficial oxygen, which fauna need to exist). Since the massive deaths of bees began a few years ago, possibly caused by a microbe whose existence is enhanced or made more extensive because of human-caused or -exacerbated and -accelerated global warming, or by humans' use of chemical pesticides that are harmful to bees, people have learned the extent to which bees cross-pollinate plants that are needed by people for food. Bee extinctions would cause regional and even global hunger if they were to continue at their present pace.

Currently, global warming increases throughout Earth. Human overconsumption and overpopulation are principal contributors, particularly because of humans' release of particulate matter into the atmosphere through motor vehicles, power plants, and irresponsible industrial production. A May 10, 2013 *New York Times* article by Justin Gillis, "Heat-Trapping Gas Passes Milestone, Raising Fears," stated that data collected and analyzed at two sites established that for the first time in human history carbon dioxide (CO_2) was measured at 400 parts per million for a full twenty-four-hour period. National Oceanic and Atmospheric Administration (NOAA) scientific instruments atop the Mauna Loa volcano in Hawaii, and Scripps Institution of Oceanography instruments in San Diego, California, independently noted and reported their findings. The last time CO_2 reached that level was at least three million years ago, according to geological research. Earth had a much hotter climate (CO_2 and climate are inextricably intertwined) and the seas likely were significantly higher—by sixty to eighty feet. (Imagine if the water that meets the shore in Miami, New York, Boston, and Los Angeles, among other U.S. cities and towns, were six to eight stories higher on skyscrapers—and flooded transportation systems' tunnels.) In such a context of drastically increasing CO_2 levels, the role of trees in cleansing the air, always extremely important, is growing to be ever more so—but massive human deforestation of still existing forests and jungles continually imperils trees and strains their abilities to provide for life. So, alien humans (significantly assisted today by the fossil fuel industry, whose corporate representatives and the politicians they influence in one form or another are culpable culprits for denying the reality of global climate change) are destroying their own habitat and extincting native vegetation that could help them diminish or mitigate their irresponsible behavior. In mid-September 2014, scientists from the National Oceanic and Atmospheric Administration (NOAA) stated that June, July, and August

2014 comprised the hottest summer ever recorded on Earth, and that the year 2014 likely would end as the hottest year on record. Descendants of European immigrants in the United States have become aliens once again, as they have alienated themselves from the biota and abiotic setting native to this place.

Alien Plants

Scientists and other people regard as "noxious" those alien plants that enter an area and endanger or diminish native or other vegetation beneficial to humans or their livestock. The once-alien humans become locked in biotic battle (in which humans are sometimes assisted by abiotic chemicals that threaten the health of humans and other life) with the now-alien noxious plant immigrant aliens for a newly contested familiar place. Sometimes humans have introduced, intentionally or unintentionally, noxious weeds that previously were unknown in a place—for example, for purposes of landscaping terrain or of eliminating previously identified undesired vegetation from farms and fields, or providing a hardy, regionally new form of agricultural crop. (An unexpected consequence resulted, for example, when U.S. farmers imported seed for Russian winter wheat because it would be hardier in colder U.S. areas than native species of wheat had been. Agriculturalists did not realize that Russian thistle and other weed seeds were present in the imports.)

Numerous noxious weeds familiar to urban and rural residents are imports. Dalmatian toadflax (*linaria dalmatica*) came from Eurasia; leafy splurge (*euphorbia esula*) from Europe; Russian knapweed (*centaurea repens*) from Eurasia; Russian thistle (*salsola kali*) from Asia; and spotted knapweed (*centaurea maculosa*) from Europe.

Cheatgrass (*bromus tectorum*), so called because it invades wheat fields and destroys wheat crops, thereby cheating wheat farmers of their expected harvest and livelihood, arrived in British Columbia around 1899 and spread south deep into the United States. The Norway maple came to the United States in 1756 as an ornamental and is now pervasive on the East Coast, where it displaces sugar maples and beech trees. The ailanthus tree ("tree of heaven") arrived from France circa 1784, after being taken from China to France in 1756; it, too, displaces native trees.

Immigrant plants beneficial to people and other biota have come too, whether inadvertently or intentionally. Among these are wheat, a food staple with multiple varieties adapted to diverse climates and the extent of water available (from rain or irrigation). Bread wheat (*Triticum aestivum*) is used extensively throughout the world. Its earliest use for food has been traced back to Turkey, to at the latest during the tenth millennium BCE. Charles C. Mann in *1493: Uncovering the New World Columbus Created* states that its introduction into the Americas came by way of an Ethiopian who had taken the name Juan Garrido while traveling through Spain after escaping from slavery in Portugal. Eventually, he accompanied Hernán Cortés on his invasion of/adventures in Mexico. Garrido planted three kernels of bread wheat in Mexico to provide a food staple for the invading Spanish soldiers and colonizers. Historian Francisco López de Gómara wrote in 1552 that one of the kernels produced 180 kernels, the multiplication continued, and "little by little there was boundless wheat. . . . To a black man and slave is owed so much benefit!" Over time the wheat spread northward, resulting today in the "amber waves of grain," sung about in "America the Beautiful," that have become especially extensive in the U.S. in Midwestern and Western states.

Alien Animals

Throughout human history, humans and some animals have gradually formed collaborative relationships. The association developed, in some cases, with animals (such as dogs, and later cats) living in or near (such as cows and chickens) human dwellings for mutual benefit. The once "other" animal became, for humans, helpful and even necessary; sometimes the helper became a member of the human household.

Sometimes, too, immigrant "exotic" animals, having been introduced and adapted, remain welcome in alien territory they had not previously inhabited. People come to find them indispensable for a variety of reasons because of the contributions they make to individual, family, or community wellbeing. Over time, the symbiotic relationship is so embedded and accepted that people do not realize that a species was neither native to the territory in which they are now living, nor co-adapted to the humans with whom they relate today. Take the family dog, for example. Its ancestor millennia ago, the gray wolf, might well have attacked people or their livestock to acquire food, in predator-prey relationships. Now dogs serve as affable companions, protectors of home, business, and children, detectors of illicit drugs and hidden explosive devices, and guides for the blind, among other roles. These animal aliens, originally a feared "other," have come to be called "man's best friend." In these and other ways they have become very much a part of human communities. A species once regarded as "alien" is now taken for granted as part of human life because of its contributions to its human home and hearth, which have become its primary and, in some cases, sole habitat.

What? My Dog's an Alien?!?

Canis lupus familiaris, the domestic dog, is descended from the gray wolf (that's the *lupus* part of its name). It was probably domesticated as many as thirty-three thousand years ago, although perhaps more recently, and maybe in Siberia. The earliest U.S. remains were found in Texas, dated to at least nine thousand years ago. So all members of *canis* or *canis familiaris*, known to their humans as Sacha, Snoopy, or just plain Spot (or Pluto, in Mickey Mouse cartoons—not to be confused by the once-upon-a-time planet of the same name), were aliens, too, at some point; today, their descendants live in our homes and backyards.

. . . and My Cat, too?!?

Felis catus, the domestic cat, had its origins in Egypt, some four millennia in the past. Gradually, it too came to be accepted and welcomed by human individuals, families, and communities not only for its ability to limit mouse populations in homes and barns, but as a companion and confidant.

Dogs and cats, aliens in past human history, are now an accepted part of the human experience, in the U.S. and elsewhere throughout the world.

At Least the Cowboys' Horses Are All-American Natives

I have bad news for you. While horses were native on this continent millennia ago, perhaps when mastodons roamed North America, eventually they became extinct here. The earliest human inhabitants of the continent probably hunted them for food (similarly, a friend told me years ago, horse steaks were available for a time in the Harvard Faculty Club restaurant—but that might just be an urban legend), so the contact between species was much different than that of the legendary cowboys who have used them far differently to travel, hunt, and herd livestock, and with whom they often had a friendly relationship, and shared mutual respect.

Sorry about that: sometimes what is so familiar and part of our culture we assume has been around "forever"—or at least as long as we and our ancestors and native peoples' ancestors have been here. Current horses are descendants of Spanish explorers (who used them to facilitate travel, carry baggage, and fight against native residents) and settlers (who used them for travel, to carry cargo, to plow farms, and in battle). Subsequently, resident and native Indian peoples adopted horses for their use (travel, transport of goods—including tipis that were the original "mobile homes" in the Americas—hunting, and fighting alien intruders, i.e., Europeans and then Euroamericans, as they tried to protect their homeland communities, agricultural fields, fishing sites, and nature-integrated way of life).

Airborne Aliens

We're not talking here about human immigrants or possible future ETI immigrants. Briefly, we should at least mention aliens who arrive by "air," in one form or another. Often, but not always, these aliens are viewed negatively. They have included birds (such as English sparrows whose population exploded rapidly to the point where, without natural enemies, they displaced native songbirds; parrots and others that have brought disease to other birds and humans, including avian flu); insects (need we mention "killer bees" which originated in Africa?); and viruses, both biotic (that have caused illness and death, some of which might be deliberately sent as part of a country's or terrorist cell's biological warfare) and electronic (including some from distant countries that are "airborne" in cyberspace and the "cloud," and might disrupt military operations, commercial transactions, and individual bank accounts, and enable theft of intellectual property, military secrets, industrial plans, and personal funds). The airborne aliens, many of which are unseen to the naked eye, can prove to be the most insidious invaders because of the problems they can cause for health, safety, security, and societal wellbeing in general.

Alien People

Sometimes native residents can become so outnumbered, overwhelmed, and oppressed that they become regarded, effectively, as aliens in their own

Aliens and other Others

land. In fact, as described eloquently in the historical and cultural analyses of the diverse Indian scholars who contributed to *Exiled in the Land of the Free: Democracy, Indian Nations and the U.S. Constitution*, edited by Oren Lyons and John Mohawk, impacts of European immigrants and their descendants have been socially, physically, and culturally devastating. Both Lyons and Mohawk are Haudenosaunee, People of the Longhouse. (In 1721, when the Haudenosaunee went from a five-member to a six-member confederation, Europeans began to call them Iroquois or Six Nations.) The once sole human inhabitants of Turtle Island have been relegated by ethnocentrism, racism, greed, and religious intolerance to an "other" and even "alien" status by the much more recent immigrants who began to arrive in the fifteenth century. Indeed, Euroamericans who are descendants of ancestors who arrived as aliens little more than five centuries ago now call themselves "natives," and have come to regard themselves as superior to and privileged over the real natives of the U.S. whose ancestors arrived more than ten millennia ago.

On September 12, 2001, the day after the terrorist attack on the World Trade Center in New York City, a striking example of this was described in Montana newspapers. An elder Indian woman in her eighties was sitting on her front porch in a rocker. As she peacefully rocked back and forth, observing life on the sidewalk and street, keeping to herself except to greet friends going by, a passing pickup braked abruptly to a screeching stop. The driver of the truck stuck his head out the window and yelled, punctuating his words with a few expletives, "Go back where you came from!," and sped away with a squeal of tires, raising a cloud of dust. (Talk about racial profiling!) The woman was surprised and angered. After all, she was a lifetime Montanan, and for thousands of years before she was born in the relatively recent U.S. state (Montana became a state in 1889), itself relatively recently named, her ancestors had lived and hunted and roamed throughout the territory. Then she regained her sense of humor, and later reflected with a wry smile as she said to a reporter, "I should've yelled, 'Go back where *you* came from! I been here ten thousand years!'" Indeed. She was the native "American." He was the recent immigrant, still alien to the land whose native places and people and their histories were unknown to him.

The truck driver's attitude reflected, however subconsciously and unknowingly, the attitude of his European ancestors when they arrived as aliens on the shores of Turtle Island on a sacred Mother Earth. The European nonnatives were discovered by the natives first as occupants of sailing ships, and then as occupiers and usurpers of native territories and robbers of natural goods. Europeans rationalized for themselves actions that were illegal by European law of the time, and violated, as they themselves knew, the Natural Law that philosophers, theologians, and clergy at all hierarchical levels discussed and taught (going back to Thomas Aquinas in the thirteenth century) and kings understood and sought to obey, as counselled by the preceding professionals. (The European Discovery Doctrine, also called the Christian Discovery Doctrine because popes and other religious leaders in the fifteenth century and afterward used Christian doctrine to justify expansion and imposition of their "true religion," is elaborated extensively in *Cosmic Commons*.)

The term *alien* has not been applied to the English, French, Dutch, Portuguese, and Spanish explorers and colonizers, and their accompanying clergy and soldiers. For Indians throughout the Americas for centuries and in the present, however, Europeans, African slaves, and immigrants have been aliens. The U.S. government has subtly enforced Discovery Doctrine here, too, by categorizing newcomers, particularly the undocumented, as "aliens"—even if they were raised since infancy in the U.S. In Indian eyes, those Europeans who arrived in the Americas from the fifteenth century to the seventeenth century were indeed aliens, and very powerful ones who invaded to take Indian lands and natural goods. (The sci-fi film *Avatar*, discussed later, projects such conduct into the future while simultaneously providing a social commentary on past and present aliens' injustices perpetrated upon native peoples.)

I'm a Native: Dad's a Son of the Alamo, and Mom's a Daughter of the Revolution

Sigh. I hate to be continually a bearer of bad tidings, but you're a relative newcomer. European immigrants arrived as settlers only in the fifteenth century CE, tens of thousands of years after the first human natives. Many of the native peoples and their communities in the Americas are theorized by European and Euroamerican anthropologists and historians to have come sixty thousand or more years before Vikings or other Europeans arrived; they were immigrants and settlers who came from Asia via a theorized "land bridge" or "ice bridge" and seafaring craft, and settled on coastal areas in what is now called Alaska, Canada, and California in North America, and Chile and other countries in South America. However, Indian historians and spiritual leaders contest the European and Euroamerican foreigners' speculation and assessment: their oral history traditions, including spiritual narratives, describe their emergence on Turtle Island (now called North America) itself, not their migration from Asia.

Aliens in the Bible

The Bible—both the Hebrew Scriptures and the Christian Scriptures—describes and discusses the "alien" and "aliens" in numerous passages. In fact, in the New Revised Standard Version of the Christian Bible (the most widely used translation), *alien* is used seventy-five times (only once in the Christian Scriptures, referring to Moses in Midian), and *aliens* occurs fifty-nine times (four times in the Christian Scriptures). In the Hebrew Scriptures the terms ordinarily refer to persons and people who are not ethnically Jewish; however, Jacob and other revered figures are called "alien" when they are living outside of their homeland. Perhaps the most provocative use of aliens occurs in Leviticus 25:23. The passage describes Yahweh setting requirements for the Jubilee Year, in which lands are to be redistributed to prevent a landed aristocracy from owning and ruling over all Israel. Yahweh states emphatically that the lands into which the Hebrew slaves flee after escaping from Egypt are God's: "The land shall not be sold in perpetuity, for the land is mine; with me you are but *aliens* and tenants" (emphasis added). Ultimately, in the Bible, God the Creator is the "owner" of the whole Earth. All humankind is "alien" to Earth in that no nation or person can claim that they own absolutely and perpetually, as national or personal property, the territory on which they reside.

Different versions of English translations of biblical texts use synonyms whose collective consideration helps clarify the extent and limitations of those who are "other" than us: foreigner, stranger, alien. Each of these carries the notion that the persons about whom we are speaking are not native to our place. In common usage we might understand, beyond the basic meaning of "nonnative" shared by all three, specific characteristics not apparently overtly shared by the others.

Foreigner designates a person from another country. They might be unknown to us personally, but we see them or hear about them. They might be known, at least superficially and at times more personally, as a resident of our neighborhood who has come into our own community to live and work among us temporarily or permanently (and at times, in either case, often depending on our particular prejudices, always might be regarded as a "foreigner," even if they become a naturalized U.S. citizen). They might be known more in-depth because they are neighbors with whom we associate because of our children's shared athletic or dance activities, our congenial conversations during backyard barbecues, or because we work with them when our employer desires their particular skills.

A *stranger* need not necessarily come from a foreign country; in fact, most often they do not: they might be walking or driving through our neighborhood. If we are particularly concerned about our children, having heard about abductions, we might be especially watchful when we note their approach or presence; if we are concerned about our property—homes and their contents, or motor vehicles—we might weigh mentally whether or not they seem to be looking to add to their possessions at our expense.

We tend to view as *aliens* those nonnatives who have some attributes apparently significantly incongruent with our own: there's something about them that makes us particularly uneasy because they are so distinct from us. They might or might not be foreigners, but they seem to be something more, whether specifically definable or unnervingly undefinable. For these reasons, we call someone an alien if they are, in our judgment, very unlike who we are. No one would be more alien to us than someone who is not only not from our neighborhood or our country, but also our planet—and perhaps even our solar system and galaxy. "Aliens from Mars" are far more threatening, at least upon first Contact, than are "aliens from Mali." We can at least do some research on the latter, and find that they are ordinarily very peaceful people; we cannot do like research about Martians.

In the Bible, when diverse "aliens" (or "strangers" or "foreigners") are noted and discussed, the presence of some might be desirable, but that of others undesirable. Some are admiringly accepted; others are culturally rejected, even when family bonds link them to the Israelites. Those accepted and esteemed include Ruth, a Moabitess; the Ninevites who repented when the prophet Jonah preached to them; Job from the land of Uz, the territory of the Edomites; and the Good Samaritan in Jesus' parable of the same name. Those rejected included the "sons of god" who had intercourse with the "daughters of men" in Genesis; and the foreigners who married Israelites—and their children—who were in Israel during the time of Ezra and Nehemiah.

Aliens and other Others

Accepted Biblical Aliens

Ruth, whose story is narrated in a book of the same name, was a Moabite. Her husband, an Ephrathite, had moved from Israel to Moab with his parents and brothers. Both brothers had married Moabite women, Ruth and Orpah, after their father died, but then the brothers died too, unexpectedly. Naomi, their mother, decided to return to her homeland. Orpah stayed in her country, but Ruth insisted on accompanying her mother-in-law to safeguard her on her journey home. When Naomi pressed her to stay with her people, Ruth responded, "Where you will go, I will go; where you lodge, I will lodge; your people shall be my people, and your God my God." Ruth left her homeland, where she was a native Moabite, and went with Naomi, an alien in Moab, to Bethlehem. There, their status was reversed: Ruth became an alien, and Naomi was restored to being a native. Ruth went to the fields of Boaz, a kinsman of Naomi's husband, to glean during the barley harvest then underway. Boaz and Ruth married soon thereafter. She would always be "alien" ethnically, but ceased to be an alien culturally or in family relationships. Ruth and Boaz had a son, Obed. He had a son, Jesse, who later was the father of David, who would become the most revered ruler in Israel. Ruth, an alien and foreigner, was the great-grandmother of King David. There's more to the story: the first chapter of Matthew's Gospel endeavors to trace the genealogy of Jesus. Matthew notes in the first verse that Jesus is the "son of David." He goes on to list David's descendants (solely in the male line, as was customary at the time), and after generations have passed he comes to "Jacob the father of Joseph the husband of Mary, of whom Jesus was born, who is called the Messiah." Ruth, an alien remembered for her loving self-sacrifice to care for her mother-in-law, is a direct ancestor of King David and Mary's husband, Joseph, and an indirect ancestor of Jesus since Joseph, his stepfather, was married to Mary.

In the well-known story of Jonah, the prophet is told to preach to the Ninevites and tell them to change their ways. Since Nineveh was a major city and one of the capitals of the Assyrian Empire, and the Assyrians were a fierce people who extended their empire and incorporated into it, by force, northern Israel, Jonah was understandably fearful of these foreigners—a palpable and contextually justifiable focused xenophobia—because he expected them to kill him, an alien representative of an alien God—perhaps slowly and torturously as evidenced by their prior conduct toward their enemies. Jonah fled in a boat going in the opposite direction. A fierce storm came up and Jonah was tossed overboard by the sailors when he instructed them to do so: he told them that the dangerous storm came because God was upset with him alone. After he was saved by a "big fish" and carried within it to the shore near Nineveh, Jonah "disembarked" and nervously preached repentance to the Ninevites. He was amazed and angered when the alien (to him; they were in their own nation) Ninevites, unlike the Israelites when the prophets preached to them, repented of what they had done to offend the God of Israel, whom they would have viewed as an alien deity. Jonah was upset because, like a good nationalist and ethnocentrist, he had hoped that they would reject his message and that then God would rain down destruction on the enemies of Jonah's people. The Ninevites, as an exception to how Assyrians are ordinarily viewed in the Bible, are recognized and respected in this story for having listened to and obeyed the words of a foreign—alien—Israelite prophet who was a messenger of an alien divinity. It is not an unambiguous acceptance: in every other instance, the Assyrians are described as invaders who seek and eventually are successful, in acquiring the land of Israel and incorporating land and people into their empire. They also relocated peoples whose lands they acquired: their enemies could be more easily controlled in terrain unfamiliar to them, where the territorial gods were unknown to them. Later in the biblical historical narrative, the Babylonians would use the same tactic when they conquered Assyria and then Judah (the southern part of Jewish homelands that the Assyrians had been unable to conquer), destroyed much of Jerusalem, and forced the people into a Babylonian exile. The exiled Jews became aliens in aliens' alien land. (Modern versions of the forced relocation tactic were evident in the U.S. in the East when the newly independent states' federal government used coerced relocation to drive Indian peoples first to the West, beyond the Mississippi, and then, for many natives, to permanent residence in the "Indian Territory," most of which is now in the South, in the state of Oklahoma; and in South Vietnam in the "Phoenix"/"strategic hamlets" program where Vietnamese people viewed as hostile were moved into relocation camps that were surrounded by fences guarded by U.S. and South Vietnamese troops.)

Job, whose name also serves as the title of a biblical book, lived in the "land of Uz," probably Edom. The Edomites were a people whose land bordered Israel; battles and wars erupted periodically between the two peoples. Job, then, was not only "alien" to Israel, but also an "enemy" in terms of historical conflicts between the two peoples. Job's identity as an "other" is not discussed in the book, however. Job's patience in the face of extraordinary adversity, and his enduring faithfulness to God, commended him to biblical writers. The regard for Job continued through millennia among many peoples as they reflected on his virtues. His remaining steadfast and faithful despite the turmoil around him, including the manner in which his "friends" disparage him, has been viewed as exemplary over the ages. In terms of our current discussion, he is regarded as a "good alien." His virtues make him acceptable, even to the extent of being cited specifically as an alien and yet being included in the Bible with an entire biblical book named after him that recounts his tribulations and his responses to them. He is often presented as an example of one who experiences undeserved suffering and evil. The question of why God allows this and does not rescue him illustrates the theological and moral issue of *theodicy*: how can a just and powerful God permit the suffering of innocents? The question is not really answered in the story. The story does present another profound theological point: the just do not always triumph; the wicked sometimes prevail. And another: just because someone suffers does not imply that they've done something wrong. Illness, accidents, and other afflictions were viewed, when the book was written (and in Jesus' time—he rejects such an idea—and into the present), as divine punishment for sinfulness: one's own, or even a parent's or grandparent's. The Job story rejects this type of thinking, which was an early and unique type of "blaming the victim."

The story of the Good Samaritan is one of the most well-known of Jesus' parables. After Jesus instructs people gathered before him that love of God and love of neighbor comprise the Great Commandment, a Pharisee adversary queries "Who is my neighbor?" In reply, Jesus recounts the parable. It is important to note that the people of Samaria and the people of Israel disliked each other—even though they were related by blood, and each had a temple dedicated to Yahweh. In the story, after an Israelite and a Levite (a person dedicated to Yahweh who worked in the temple) pass by a fellow Israelite who had been attacked by robbers and left for dead, a Samaritan "alien" stops, attends to the immediate needs of the "enemy alien" Israelite, takes him to the nearby town, and provides money for his lodging during the time he is healing. The Samaritan then tells the innkeeper to whom he has paid for food and lodging to enable the Israelite to recover from his wounds, that on his return from his business journey he will stop and pay for whatever other expenses were incurred to pay for the needs of the grievously wounded Israelite. In this story, the alien Samaritan passerby exemplifies *xenophilia* (also, in Greek, called *philoxenia*), love of the stranger/foreigner/alien—who is also viewed as an "enemy."

In the Gospel according to John, a story about Jesus taking a shortcut through Samaria during one of his journeys describes a Samaritan who is compassionate toward a Jew—in this case, a Samaritan woman who encounters Jesus. After she gives him water from the community well and has a short exchange with him she becomes, effectively, an early missionary of Jesus. She runs to her village to relate her experiences to her people, among whom she had been alienated as an outcast. In the narrative, when he meets the "woman at the well" Jesus violates several social conventions and religious legal requirements: while alone, he speaks to a solitary woman; the woman is a Samaritan, and he would become ritually "unclean" by being near her and, even worse, to ask her for a drink from a vessel from which she had sipped, which rendered it "unclean" for Jewish lips; and then instructs her in the message he has been preaching in Israel. The woman says that it is odd that he, a Jew, would speak with her, a Samaritan. After she tells the people of her village, several go to the well out of curiosity, and Jesus instructs them, "enemy to enemies" in the belief to the time. Like Jonah, Jesus gives a message to an "enemy," and like the repentant Ninevites who listen to and obey Jonah, the woman and members of her community believe in Jesus and his words, unlike some of Jesus' fellow Jews whom he had instructed.

Rejected Biblical Aliens

Genesis (6:1–4) has a very brief narrative about a time when, as the human population increased and expanded on Earth, "sons of God" from the heavens found the "daughters of humans" very attractive, "took wives for themselves of all that they chose," and had intercourse with the women, who bore their mutually conceived children. The latter offspring of divine-human (alien-human?) unions were the Nephilim, who were very tall and powerful. (Some UFO researchers and devotees regard this incident as a biblical narrative describing ETI who came to Earth and planted their biological seeds among women who had a biblically and culturally forbidden divine-human sexual encounter.) The biblical story of a global flood follows then, implying that one of the human sins that provoked its punishment was the intercourse between "sons of divine beings" and "daughters of humans" whose progeny were a mix of both exoEarth and Earth intelligent beings. The incident also provokes God to shorten and limit the human life span to a mere 120 years. Earlier ages for ancient ancestors were in the hundreds of years. Noah, for example, whose story begins before and ends after the flood, is said to have lived to nine hundred fifty years.

The type of religion-state union of which many thinking people (who are not adherents of the "true religion" or are not of the same ethnicity of those who become dominant and oppressive) are rightfully fearful today is described in the books of First and Second Chronicles, Ezra, and Nehemiah. A Persian-appointed governor, Nehemiah, and a priest and scribe, Ezra, use their power—concurrently, consecutively, or in different periods (the history is debated)—to purge from Israel any foreign blood—spouses and children—who are polluting the ethnic purity of the Jewish people. (Some would regard this as an early form of "ethnic cleansing.") Those to be alienated from the lands of Israel are the foreigners who have crossed ethnic and cultural lines and made a life together with the Jews. They had not accepted previously operative cultural condemnations regarding the "other," but subordinated these to mutual love or needs. But not only those who were clearly aliens were exiled from Israel; those who were descendants of aliens were similarly treated. Thus, wives whose husbands were full blooded Israelites, and the children who were their offspring, were forced to be separated from the man who was the husband and father whom they loved and who provided for their sustenance. (Ruth, the alien Moabite and ancestor of King David, would not have been exempt from being regarded as an alien had this occurred during her lifetime. She would have been separated from her husband Boaz, and sent into exile—with their children. (A contemporary parallel in the U.S. is that some opponents of U.S. immigration reform want to expel from the U.S. all children who were brought into the country illegally when they were infants carried by illegal immigrant parents. It makes no difference if the parents have been hard-working, productive residents; neither does it matter if the children have grown up here, graduated from high school and perhaps college and have been productive, tax-paying residents. All of these "aliens," these "illegals," would be deported without sympathy or compassion.)

Biblical writers, lawgivers, and prophets express concern for aliens throughout the Hebrew Scriptures. They know that the ancient Hebrews were oppressed when they were aliens in Egypt in the past, and because of their experience their descendants should act differently and be compassionate to aliens in the present rather than do to others what had been done to them. Leviticus 19:34 expresses this well. Yahweh instructs Moses thus: "When an alien resides with you in your land, you shall not oppress the alien. The alien who resides with you shall be to you as the citizen among you; you shall love the alien as yourself, for you were aliens in the land of Egypt: I am the Lord your God." Centuries later Jeremiah is commissioned by Yahweh to tell the people going to the temple in Jerusalem that Yahweh does not dwell there any longer because they act unjustly toward each other, worship false gods, and "oppress the alien, the orphan, and the widow"; when they stop acting in this way toward the most vulnerable in their community, Yahweh declares, "then I will dwell with you in this place" (Jer 7:5–7).

In like manner Jesus teaches in the Last Judgment story (Matt 25:31–46) that solicitude for a stranger, expressed in hospitality to aliens in need, will be a criterion by which people will be judged when they stand before the divine judge hoping to be invited into the reign of God: "I was a stranger and you welcomed me." Jesus teaches, too, complementary to Yahweh's commandment to Moses to "love the alien as yourself," that the second part of the Great Commandment is to "love your neighbor as yourself"—and for Jesus, all peoples, including alien Samaritans in the parable, are "neighbors."

Suspicious Strangers Sighted

Since earliest human history, a newcomer coming into a nomadic or resident community's territory would be regarded with suspicion, fear, and hostility. Among many peoples, there was no capital punishment for an egregious offense or harm against a member of their own community: "Do not kill" was meant for any and all circumstances. People did not want on their hands the blood of a relative or someone else in their community. So, a murderer would be sent into exile, under penalty of death (a last resort) if they returned. Exile was considered capital punishment because it would be extremely difficult for a solitary individual to survive on their own. They would need the skills to hunt for food, to avoid being food for carnivorous predators, to make their own hunting equipment (spears, bows and arrows, etc.) after finding needed materials to do so, to provide clothing, to erect shelter, to survive the vagaries of weather, and to negotiate their conduct and interaction with other humans whom they would encounter.

Aliens and other Others

Among some peoples, because of the preceding, outsiders were killed on sight on site—the original "shoot first and ask questions later" policy and practice. Native inhabitants assumed that the new arrival was probably a scout for an invading party, or a sentenced murderer. If the former, the scout posed a threat to the whole community, for he (rarely: she) would report back that natives existed in a place, where they lived and how extensive their community appeared to be, the kinds of land and water they had, the natives' defense system if any, number of warriors and their weapons, best time to attack, and anything else that would help incomers be successful invaders. If the newcomer were perceived to be a murderer, the natives would judge that if someone did something so terrible to a member or members of their kin community that they were exiled, how much worse their conduct would be toward members of a community where they had no like relationship. In either case, then, the potentially dangerous exile had to be killed to prevent them from perpetrating some evil against native people(s), and the presumed scout must be killed to prevent them from conveying community information back to their invading companions. The approaching stranger was in danger once discovered: those who were assumed to be criminals or scouts were killed outright to protect the community against anticipated individual or military predation against people, property, and place.

The biblical story of Cain and Abel in Genesis 4 illustrates well the practice of exiling a murderer, and the fear the murderer would have that the people encountered would kill him, because for their part the resident people would fear him as a dangerous, perhaps homicidal, interloper. In the story (a mythical narrative that teaches religious and moral doctrines, and is not meant to be read as objective history), the first members of the human species, Adam ("from the earth") and Eve ("mother of all the living") have intercourse. Eve conceives their firstborn, a son whom they name Cain. Eve conceives next a second son, whom they name Abel. Cain becomes a farmer, and Abel a shepherd. When they offer to God what their labors have produced—Cain offers the "fruit of the ground" and Abel "the firstlings of his flock, their fat portions" (which indicates a premium offering)—God looks favorably only upon Abel's offering. This makes Cain angry, and he murders his brother. God punishes him, sentencing him at first to a difficult life in his occupation as a farmer, since Earth will be difficult for him to till and will not be as productive as before; and second, to be in exile from his native land and his original agricultural fields: he is to be "a fugitive and a wanderer on the earth." Cain replies, "My punishment is greater than I can bear! Today you have driven me away from the soil, and I shall be hidden from your face; I shall be a fugitive and a wanderer on the earth, and anyone who meets me may kill me." Now, the perceptive reader might wonder, "Wait a minute. Adam and Eve are the first humans, and Cain and Abel are their only children at this point in human history: why is Cain fearful that someone will kill him, if there are no other people in existence?" The answer is that in order to highlight the gravity of the sin and crime of murder, the biblical writer "retrofits" a practice from a much later historical era, in which a murderer is sentenced to exile and is greeted with fear and hostility by natives noting the arrival of the alien in their territory.

In Jesus' teachings, as discussed earlier, the strongest and most direct reference to a required Christian attitude of love of the other in need, contrary to conventional practice, is stated in the Last Judgment story and reinforced in the Good Samaritan parable. A consciousness and conduct of welcoming and treating well strangers and aliens is taught also in the Torah and by the prophets of ancient Israel, in the Holy Qur'an, and in other religious traditions, and exemplified in the social-spiritual practices of many indigenous peoples throughout the world (think, for example, of the Pilgrims in Massachusetts who couldn't have survived their first winter ashore without the help of the Wampanoag or Pawkunnakut native peoples).

What's All of This Have to Do with Encountering ETI?

Remember Stephen Hawking's comments in the introduction to our *ETI* exploration? He said that the radio signals transmitted by the SETI (Search for Extraterrestrial Intelligence) Institute should be terminated, and all efforts to Contact ETI should cease: we don't want aliens to be attracted to our planet to pillage and destroy. He projects ETI conduct on Earth to be intrusive and invasive just as Europeans' conduct had been when they arrived in what they would call the Americas from the fifteenth century on. He declared that Columbus' arrival "didn't turn out very well for the Native Americans." So, Hawking states, we should see just what that European conduct was, and what sort of consciousness European explorers and colonizers brought with them to rationalize their imperial, illegal (by European law and natural law as they understood it) acts, and their consequent unjust behavior toward native peoples who had cities, farms, hunting areas, and sacred sites, were integrated with their Earth context, and recognized

Aliens and other Others

their interdependent relationships with other biota, all of this while living in communal societies without private property. Discovery Doctrine (which asserted that Europeans had a civilization superior to all others in the world, and had the "true religion") was what European countries asserted to justify their conduct toward native inhabitants, such that things "didn't turn out very well" for them.

A political and socioecological danger lurks: the ubiquitous use of Discovery on Earth continues today, enshrined in national and international laws. In years to come, Discovery impacts might well, barring ideological transformation, color the activities and claims of individual nations and humankind as a whole during space exploration (even through telescopes) and eventual colonization. Intelligent beings encountered in outer space places who have organized social arrangements, residential and commercial structures, and established agricultural operations, whether seen from afar or close at hand might be judged outright as "inferior" aliens by human alien explorers who claimed to express the hopes and ideology of humankind (or at least of the industrial, military, and political leaders who fund their space voyages in a hunt for new territory to colonize and natural goods to exploit).

Contact in the "New World": Who's the Alien? Who Discovered Whom?

Let's explore Discovery Doctrine further. We'll take an intellectual journey back to the European Age of Discovery (fifteenth to seventeenth centuries) initiated by Portugal and Spain, the first nations to develop long distance seafaring capabilities. The explorations of these monarchic rivals encountered territories and peoples previously unknown. Soon, other European nations including England, France, Holland, and Sweden became similarly technologically developed, and the race was on for territory on distant shores, across the oceans and beyond the boundaries of Europe.

In order to have a self-satisfying religious and supposed legal basis for imperialism, European rulers (including popes, who ruled over extensive land holdings possessed by the Catholic Church) and their advisors devised a Discovery Doctrine, later called by some historians a Christian Discovery Doctrine, and often shortened in discussion to solely "Discovery" (always capitalized to distinguish it from ordinary and usual uses of the word *discovery*). Robert J. Miller, Eastern Shawnee, legal scholar and professor of law, provides an extensive and in-depth account of Discovery in *Native America, Discovered and Conquered: Thomas Jefferson, Lewis and Clark, and Manifest Destiny* and in his contribution to the volume *Discovering Indigenous Lands: The Doctrine of Discovery in the English Colonies*. He discusses how, developed and strengthened over time (from the fifteenth century to the twenty-first, and still being used), Discovery has been instrumental in depriving native peoples of their lands, cultures, religions, and rights. Its impacts have spread globally. In particular, Western nations beyond the U.S. have appropriated its "principles" for their own nationalistic purposes.

During the Discovery period European nations came to be more externally defined in relation to each other, and internally cohesive, if not entirely unified. Their shared cultural characteristics, particularly their monarchical political systems, private property-based economic systems, and dominant Christian religion (which was understood to be the only "true faith"), came to instill in them an arrogant and assertive sense of civilizational superiority over all other nations and peoples—those already known through travel and trade, and those to be found elsewhere in the world.

Europeans claimed that their expansion to and seizure of new (to them, not to the existing inhabitants) worlds and territories was done for other peoples' material and spiritual wellbeing—as judged, decided, and enforced by the new arrivals. Europeans' rationalization was that they were politically and economically "civilizing" all other peoples in this life (for their cultural benefit and material wellbeing in the visible world), and religiously "saving" them for eternal life (for their spiritual benefit, everlasting, in a world unseen).

Historical Origins and Development of Terrestrial Discovery Doctrine

In the late fifteenth century, Portuguese and Spanish explorers, after ranging far in their vessels, upon their return declared that they had "discovered" a "New World" across the Atlantic Ocean. From that time forward European rulers, and the explorers and colonizers that they sent to find and settle "discovered" lands (which were also "empty lands" by European definition—empty, that is, of a Europe-like civilization and the "true faith"), engaged in a twofold assault upon indigenous peoples and the territories or nations they inhabited around the world.

Aliens and other Others

First, they perpetrated an "American Holocaust," in the words of David E. Stannard in *American Holocaust* and Ward Churchill in *A Little Matter of Genocide*, against those they came to call generically "Indians" without honoring any distinctions among peoples of resident native nations. Stannard states that "the destruction of the Indians of the Americas was, far and away, the most massive act of genocide in the history of the world." Churchill concurs, noting that the "American holocaust was and remains unparalleled, both in terms of its magnitude and the degree to which its goals were met, and in terms of the extent to which its ferocity was sustained over time by not one but several participating groups.... All told, it is probable that more than one hundred million native people were 'eliminated' in the course of Europe's ongoing 'civilization' of the western hemisphere."

Second, they promoted devastation of Earth's water, air, and earth and spread around the globe, including through their descendants, their environment-destructive agricultural, forestry, and property relationships, and the commercial practices that they had developed in their homeland.

Resident native peoples had villages, agricultural fields, and community-based fishing operations. Although these native peoples simultaneously discovered Europeans arriving on their territorial shores, there was not a balance of discovering, each of the other. Europeans were culturally and assertively acquisitive; more aggressive ideologically; and had developed a superior military technology and a new ideology related to the word *discovery*: the Discovery Doctrine. This ideology-infused concept meant far more than merely coming upon something hitherto unknown. Its tenets included European assertions that their monarchs had rights to ownership and use of, and exclusive commercial rights over, the territories and natural goods in lands newly found by explorers whom they financed, and who flew their particular sovereign flag.

When they became settled immigrants on Indian lands, Europeans began to carve up the territory into parcels of property "owned" by their respective monarchs or by colonies chartered by them. The newcomers waged war upon indigenous populations when it became the only means by which Europeans could seize lands from natives who did not want to part with their homelands and traditional hunting, agricultural, and spiritually significant places.

The "legitimation" of Discovery was based originally on papal authority. Steven Newcomb, Shawnee/Lenape, in *Pagans in the Promised Land:*

Decoding the Doctrine of Christian Discovery, ascribes Discovery Doctrine's origins to the Catholic Church and calls the ideology *Christian* Discovery.

A new "international law" (in reality, Europeans' laws) began to be developed to provide justification for seizure of foreign lands not ruled by "Christian princes." Initially, it integrated existing European laws that had been developed to avoid conflicts among European nations competing for "discovered lands" that had desirable territory or desired natural goods.

Indigenous peoples did not have and did not need, in their eyes, a papal or any other religious stamp of approval to occupy and use their traditional territories. In some areas, people whose ancestors had arrived a dozen millennia or more earlier than the Europeans were enslaved by the new arrivals, forced from their territories, or killed outright in order to take by force their lands and whatever Earth natural goods were located on them, which included fertile soil, virgin forests, clean water, and beneficial minerals (especially, in the colonial era, gold and silver; in later times, coal, oil, and uranium). For the Europeans, Indians did not have the cultural requirements to be accepted as "civilized": a European-like political-economic system, especially in terms of possession of private property whose ownership was delineated by human owner-placed boundaries such as fences; and, the Christian religion. Native peoples needed to have, in European eyes, both the "true civilization" and the "true religion," as these had been defined by invading European colonizers, to retain their respective indigenous cultures and lands.

Where both Europe-mirroring culture *and* the Christian religion were not present, the lands found were declared *terra nullius* ("vacant land"), or *vacuum domicilium* ("vacant home")—despite existing populations of native inhabitants, or the extent of their agricultural and urban development. Europeans invoked these Discovery provisions to declare that lands were empty and their supposed ownership had not been validly established by residents (not even by virtue of occupancy for millennia and by constructed cities). In a rationalization and justification not hitherto expressed or employed in human history, popes and kings and their religiously, ethnically, and culturally biased advisors stated that "empty" or "vacant" did not mean that no one was dwelling on the land, but that indigenous peoples were not using or working their territory in the manner of Europeans. Indians were not in compliance with European customs and laws (no matter that this was not possible, since cultures around the world developed over time their own laws and understanding of property relations, if any, and responsible use) or faithful to Christianity (even if Christians, let alone Christian missionaries, had never before come into contact with natives, which would have provided at least an opportunity for them to accept or reject Christianity). Absent either or both of these requirements—a Europe-mimicking culture and Christianity—peoples encountered were subjects of and subjected to European rulers, and their lands were open to possession and use by the first European nation to come upon them.

In Indian thought and practice, however, land, water, and air could not be owned—the latter had intrinsic value that should be respected and could be negated only when people had to provide for their needs: necessity transformed *intrinsic value* (value in itself) into *instrumental value* (value for others' uses, either in its natural form or as transformed by human labor) in that particular place and time. Integrated relationally with Earth, native peoples farmed communal agricultural fields, hunted collaboratively in forests, and fished waters communally. The Earth and other natural goods were regarded as a sacred provider, Mother Earth, whose holistic integrity was to be respected. Earth was never private "property" that could be parceled, and bought and sold. Natives respected each other's territorial sovereignty, usually defined by a village's or extended family's uses of an area to provide for material needs. Natural goods converted by human labor to individual goods were regarded as the personal property of the individual—but these too were shared with others at particular moments to meet others' specific needs, because cultural attitudes of mutuality and reciprocity prevailed. The native peoples' ideal and practices were to "share all things in common," much like the early Christian community in Jerusalem founded by the apostles most closely associated with Jesus (described in Acts 2, 4).

In *Changes in the Land: Indians, Colonists, and the Ecology of New England*, William Cronon describes conflicting Indian and European attitudes, ideologies, and practices regarding land occupancy and use. For the colonists, land that was not enclosed by fences and altered for a particular use was not owned, and was therefore available to be claimed and taken as private property. The boundary-forming and property-designating and dividing fence symbolized that a place had been "improved." By contrast, the Indians' lives and practices were intimately related to place because of their primarily spatial worldview (as discussed in depth by Osage Nation scholar and theologian George "Tink" Tinker in *American Indian Liberation: A Theology of Sovereignty*). Some lived in seasonal mobility between agricultural and hunting areas; others stayed in a place but used practices such as periodically burning forests to create new habitats for diverse free (not domesticated by humans) creatures (such as deer, elk, and bear) that were especially needed periodically, and then hunted for their instrumental value (e.g., food and clothing). The colonists did not view this as Indians' exercise of control over land, since such practices were done by and for the community as a whole, not for some individual commercial benefit that would have been indicated by demarcation, particularly with fences, into private property.

Ironically, in their "new world" English colonists acted with the same callousness toward native inhabitants that they or their forebears had experienced under the British nobility, which passed a series of Inclosure Acts to remove peasant farmers from their homes, fields, and places of livelihood on nobles' large estates, where for generations they had lived, worked, and paid rent. The new arrangement generated greater wealth for the idle nobility (they replaced people with sheep), and great suffering for the once hard-working poor—because of their sudden joblessness, homelessness, and consequent great poverty and inability to provide by their own labor life's necessities. The English colonists were "doing unto others" what they had *not* wanted "done unto them."

In Discovery theory and practices, a "law" that was really a provincial law, for all its claims to be "international" law, was declared to be a global law so as to privilege Europeans and European customs and culture over wherever, whenever, and whatever native peoples' cultures were encountered. States Newcomb, "On the basis of a biblical viewpoint that the chosen people are providentially assigned the task of subduing the earth and exercising dominion over all living things, the Christians considered themselves to be chosen people divinely obligated to 'save' the heathen nations by subjugating them, euphemistically referred to as 'civilizing' them. . . . The heathens are destined by God to be saved and *reduced* to Christian European 'civilization.'"

In the U.S. after independence from Europe, a key legal factor in fostering an internal, U.S. application of the European discovery doctrine, specifically against Indians, was the case *Johnson v. M'Intosh*, settled in 1823 by the U.S. Supreme Court. The court said that Discovery was part of U.S. law because it had been operative previously as English law for the English colonies.

Aliens and other Others

The court arbitrarily and unilaterally determined that the benefit of Discovery to native inhabitants was that they were forced to change their own cultures, and religious beliefs and rituals, for those of the invaders who had murdered their people and stolen their lands. The court also employed the concept and practice of *terra nullius*—"empty land"—to reach its decision: empty, that is, of a European-like culture and the Christian religion. *Johnson v. M'Intosh* introduced also a new twist to European Discovery Doctrine. Historically in Europe when one nation conquered another, ordinarily the conquered peoples' private property in land was not taken: they kept their individual or family titles, but were expected to obey their new government. However, in the U.S., the court stated, as it added a second meaning to "conquest," this did not apply because of the conflict of American and Indian cultures and religions, and the Indians' "savagery." The court thereby broadened the concept, terms, and reality of "conquest" centuries after "discovery" and "conquest" originally had been devised and developed.

A demonstration of how arrogant and self-serving the Doctrine of Discovery was and is may be discerned through consideration of a thought experiment in which the "discoverers" are reversed. Imagine that before Europeans had left home on voyages of discovery, Indians or other peoples indigenous to Turtle Island had developed sailing ships and weapons superior to those in Europe, and traveled east across the Atlantic Ocean to England. If indigenous peoples had had a "Doctrine of Discovery" it would have asserted that a superior civilization was one in which, first, the wellbeing of the community as a whole—which includes all individuals—takes precedence over its individual members (who already share in community wellbeing); second, that equitable sharing of Earth's natural goods ("resources") was a defining characteristic of civilization; and third, that no one could claim private ownership of Earth's air, waters, or earth—these were the Creator's creation, to be shared by all biota, who are interrelated and interdependent; and that these understandings originated in the principles and natural laws given by the Creator to native peoples—to which they were faithful, as evident in their spiritual consciousness and contextual conduct. Consequently, indigenous peoples would conclude, British civilization was inferior, and had to be subjected to the superior Indian culture and religion that was brought to Europe—for the Europeans' own good, for a global common good, and to embody transculturally the Creator's intent for and instructions to all humanity.

An objective thinker today would judge that the Indians' cultural characteristics as elaborated did indeed describe a superior civilization. In fact, their communal civilization, which had common property and no private property, was the ideal presented in philosophies that originated in Europe—by Thomas More in *Utopia* and by Plato in *The Republic* and *Laws*—and in the Christian Bible, in which Acts describes the ideal Christian community sharing its goods communally. This truly superior type of civilization was more likely to have internal peace and, if its way of life were internationally practiced, external peace as well.

Singer-songwriter Bob Dylan captured the Discovery attitude toward Indians and the ongoing, dominant U.S. attitude toward every "Other" in his 1963 song "With God on Our Side." In the second verse, Dylan sings, "Oh, the history books tell it, they tell it so well. / The cavalries charged, the Indians fell. / The cavalries charged, the Indians died. / Oh, the country was young, with God on its side." In subsequent verses, Dylan cynically states that God was on the U.S. side also in the Spanish-American War and other wars. His perhaps prescient words describe what occurs fictionally in the recent film *Avatar*, and might well occur in the future as humankind voyages to Mars and beyond.

Discovery: Past, Present, and . . . Future?

In the Americas, the European attitude toward and attempts to control newly engaged cultures was carried beyond the era of exploration and colonization into their colonies after they achieved independence through revolution. It was imposed on native peoples by the dominant political and cultural ideologies and powers in the national entities that were imposed in North, South, and Central America.

The original, unjust bases for the Discovery Doctrine—religious imperialism, political imperialism, economic imperialism, cultural imperialism, and racism—have remained unexamined and unquestioned for more than half a millennium. Discovery has been reaffirmed time and again in U.S. Supreme Court decisions issued over decades, and by U.S. laws, policies and practices toward Indians in particular. Over the centuries since Discovery was first formulated, then, tenets of the doctrine have been invoked periodically, particularly in judicial proceedings. Cases continue to be adjudicated, on issues such as land and natural goods ownership and use, solely on the basis of prior courts' decisions regarding the issue at hand. The reasoning behind prior decisions, the Discovery elements that were cited pro or con to settle a case in the past, has not been critically examined to discern whether the original court decisions were just or unjust in themselves. As a consequence, past injustices are perpetuated through time, and perpetrated on new persons or social groups—especially Indian peoples. Indians as individuals and nations continue to bear the brunt of Discovery applications.

The Discovery Doctrine in its ongoing iterations, directly or indirectly, has been also the foundation for efforts by nations to range far from their borders to seek natural goods in others' territories or, on the pretext of "national security," to take military (sometimes through mercenaries), political (sometimes through local political elites as surrogates), and economic (sometimes through powerful local wealthy elites) control of other countries. It might well become operative in space exploration and colonization, U.N. treaties and other international agreements notwithstanding.

During the Great Depression, retired Marine Brigadier General Smedley Darlington Butler (1881–1940), who had been honored as the most decorated veteran in U.S. history, pondered his accomplishments in depth. He concluded that rather than fighting for "democracy" he had been in the service of U.S. corporations and wealthy individuals. In 1931, he first delivered what came to be known as his signature speech, "War Is a Racket." It soon appeared in pamphlets and booklets that were widely distributed, and eventually in several books with the title *War Is a Racket*, still in print today—and still widely circulated. In General Butler's words: "War is just a racket.... It is conducted for the benefit of the very few at the expense of the masses.... I spent thirty-three years and four months in active military service.... I spent most of my time being a high-class muscle man for Big Business, for Wall Street and the bankers. In short, I was a racketeer, a gangster for capitalism.... I helped make Mexico, especially Tampico, safe for American oil interests in 1914. I helped make Haiti and Cuba a decent place for the National City Bank boys to collect revenues in. I helped in the raping of half a dozen Central American republics for the benefit of Wall Street.... I helped purify Nicaragua for the international banking house of Brown Brothers in 1909–12. I brought light to the Dominican Republic for American sugar interests in 1916. In China I helped to see to it that Standard Oil went its way unmolested."

Butler's words serve as a continuing caution today regarding wars (in the past and present and, potentially, in the future): determine if their primary intent is to protect people or pursue profits, and who profits most from them. His statement, unfortunately, might prove to be relevant not only on Earth, where Discovery Doctrine is presently applied, but also in the heavens, where it might be applied.

The Discovery Doctrine's past history and present use raise important questions regarding the continued use of Discovery today, in the original nations in which it was applied and beyond. These questions should prompt concern about possible extension of Discovery into space exploration and settlement, and the consequent socioecological implications and impacts of its trans-contextual, transnational, transcultural, trans-galactic, and trans-species application vis-à-vis extraterrestrial intelligent life, and on commonly sought celestial bodies that have terrigenous goods needed by ETI and desired by TI.

In the Heavens as on Earth? Will Humans *Discover* Life in Space?

As robots explore planetary or lunar surfaces, and eventually are followed by astronauts, responsible citizens in all countries must demand transparency from their political leaders to ascertain the real purposes of such expeditions—perhaps especially when a government declares that ventures are needed to "protect national security." Earth's nations have signed United Nations space exploration and colonization treaties (principal among them: the Treaty on Principles Governing the Activities of States in the Exploration and Use of Outer Space, including the Moon and Other Celestial Bodies [1967], also known as the Outer Space Treaty; the Agreement Governing the Activities of States on the Moon and Other Celestial Bodies [1979]; and the Declaration on International Cooperation in the Exploration and Use of Outer Space for the Benefit and in the Interest of All States, Taking into Particular Account the Needs of Developing Countries [1996]). These documents state that no nation may own territory or goods exclusively for itself, because everything is part of the "common heritage of [hu]mankind," and natural goods ("resources") found should benefit first the least economically developed nations and the most disadvantaged groups within all nations. A key question to be asked before and during space exploration and colonization: Is this particular voyage and settlement intended to fulfill international treaty obligations or will it benefit primarily transnational and, eventually, transgalactic corporations that will seek taxpayer-financed military accompaniment? The question will help us discover if Discovery is still undermining peoples' wellbeing. One might well wonder what the correlation might be between Butler's actual responsibilities in the field in the twentieth century and the U.S. government's publicly expressed rationale for them, and military-industrial expeditions led by Butler's military counterparts on twenty-first-century space voyages, lunar bases, and planetary colonies and the publicly expressed rationale for them.

Europeans' and Euroamericans' genocide against Indians in the Americas and near-terracide against Earth should be kept in mind during discussions of potential impacts by Earth explorers and settlers in cosmic space. If current ideologies and policies on Earth are sustained, and human consciousness and conduct remain unchanged prior to space voyages to "discover" habitable or natural goods-laden worlds, then terrestrial ecological and cultural devastation will be exported to extraterrestrial places and "alien" peoples. Merely shifting segments of humanity from being Earth inhabitants to becoming occupants of new planets or celestial bodies might bring them temporary respite from human attitudes and actions on Earth, but these would soon surface on habitable places in space, and harm them and other places just as they devastated Earth, humans' home planet. In space, the operative rationales noted, in which Discovery is embedded, could be used to justify invasion and seizure of celestial bodies, in whole or in part, notwithstanding U.N. documents' ideas, ideals, principles, and goals.

On Earth as in the Heavens? Will ETI Discover Earth and TI?

An ETI Discovery Doctrine would mean that future prospects are not very good for *homo sapiens*, human terrestrial intelligence (here and hereafter the acronym TI refers specifically to humankind, not to denigrate other intelligent biota but to limit discussion to our species so that we might focus specifically on and consider impacts that we conceivably could experience). If ETI civilizations are billions of years older than ours, and their technology overwhelmingly superior, it doesn't bode well for us if they are also belligerent and bellicose, with weaponry to support their attitude. We'd be run over. We might hope that in order to have reached a unified planetary effort to set forth as a species in the distant past they must have overcome any ethnic, class, religious or other characteristics that caused conflict on their origin planet. We might also hope that, similar to the mix of cultures in the crew of the starship *Enterprise* as presented in the classic *Star Trek* television series, there are interplanetary and interspecies "good guys (and gals)" who offset and seek to thwart the "bad guys (and gals)." They will provide assistance to us—unless, perhaps, they view us as being bad guys and gals because we are constantly at war in one part of Earth or another; poverty, homelessness, and malnourishment plague other parts, including in segments of the economically richest nations; poor health care harms poor people; and greed and an overbearing private property ideology, exemplified in an economic system such as capitalism, provide a supportive context for all of the preceding. Benevolent aliens might just move on after studying humankind, following a policy of noninterference and leaving us to deal with ETI who are just like us, all by ourselves. Indeed, the variety of types of space vehicles reported and videotaped might indicate that different intelligent species are studying us carefully, following our social and ethical progress—or the lack thereof—to see the outcome of the choices that we (or a controlling part of "we") make.

In the Heavens as on Earth? Will TI Discover Other Planets and ETI?

As the human species extends into space and colonizes celestial bodies in distant places, intelligent life in those locales will be threatened by a Doctrine of Discovery exercised by humans on a cosmic scale, unless people change their consciousness and conduct prior to venturing forth. The earlier discussion of impacts on Indians in the Americas should be remembered here, particularly in recalling that Indian peoples had *communal* lands, traditional familial agricultural sites, and territorial sovereignty, rather than private property and nonnegotiable national boundaries; and, indigenous peoples adapted themselves to Earth's regional places, as much as possible, rather than solely seek to adapt Earth places to people, as often has occurred in European practice and its ideological descendants today.

Given its current attitudes on Earth, humankind as a whole might well perpetuate and perpetrate the Discovery ideology that Europeans had when they colonized the Americas, and that Euro-Americans have today, even in the twenty first century: Earth is created for human benefit; Earth is humankind's property; the Western-based "private property" ideology and its consequent practices on territorial administration and natural goods distribution determined by individuals trump indigenous communal traditions and practices of compassion and sharing based on them. If Discovery continues unchecked, ETI on other celestial bodies would experience from human explorers the result that Stephen Hawking perceptively observed occurred when reflecting on the possibility of terrestrial-extraterrestrial intelligent life Contact: as Hawking stated, Contact "didn't turn out very well" for those "Contacted" (i.e., Indians in the Americas). Similarly, as noted, if ETI has its own Discovery Doctrine, then Contact will not "turn out very well" for humans who have been Contacted. The best hope for TI humankind and ETI species is that both are able or become able to reject any Discovery Doctrine, whatever its source. In that case, Contact should turn out well for both TI and ETI.

Acceptable, Accommodated, and Accepted Aliens

In the preceding sections, we've seen how some plants, animals, and peoples who arrived on the shores of what came to be called the U.S. by the first human aliens, who arrived from European nations and were discovered by resident natives, have over time become accepted and accommodated. The alien-now-resident aliens have responded with varying degrees of acknowledgment, accommodation with, and acceptance of native inhabitants who have been "alienated" by newcomer aliens—with or without pickup trucks.

Where prejudice (racial, ethnic, or economic) exists, arriving aliens might still be acceptable by local communities and cultures if they work at menial, dirty, dangerous, low status, and low-paying jobs that the dominant culture's people do not want to do themselves. If the immigrants do well in relating to their employers and other residents, they might become accepted—but kept at a social and sometimes residential distance. "The other side of the tracks" is literally that in some communities: railroad lines demarcate distinct population clusters and densities. Eventually, the aliens might be accommodated in ways that enable them to enjoy some minimal benefits: a place to live (at times, substandard housing: lacking waterlines, electricity, and adequate ventilation, and even garbage service; some are dilapidated and fire hazards), emergency room health care, education through high school. But they continue to be regarded and treated as "the other." In recent years, politicians and businesspeople, particularly but not solely in the state of Arizona, have been increasingly strident in their efforts to limit aliens' entry into their state, and to refuse to provide them with health care, decent housing, and other benefits, and especially vocal in denying them entry into a system of permanent residency, including by obtaining citizenship and residing without fear of deportation. Ironically, perhaps, such people often enjoy alien nutrition, such as "Mexican" or "Tex-Mex" food. Finally, the alien "other" might be regarded not with condescension or mere "tolerance," but acknowledged as a neighbor, one who belongs in the community as a valuable contributor to community life, culture, and wellbeing.

Impeding Immigrants, Imposing "English Only"

In recent years, the term *alien* has been increasingly applied to Mexicans and Latinos/as almost exclusively, despite the influx of hopeful arrivals from many European nations (as well as India, China, African countries, and others), including, on the East Coast, Irish immigrants to Boston and New York City—some residing illegally. It is, of course, easier for citizens to note how porous is the Mexican border, and increasing numbers of "foreign" Mexicans who have resided in the U.S. for years are gainfully employed. Across the U.S. their work fulfills the labor requirements of local businesses, contributes to community stability while ensuring personal and familial security, provides tax monies to meet community needs, and enriches area culture. They often are the people seen cleaning streets or collecting garbage, working on construction sites, and employed in restaurants as cooks, waiters, and dishwashers. Sometimes, racism has extended to Mexican Americans who are citizens, working people, highly family-oriented, and perhaps a second or third generation of citizens—or longer, going back to when what is now the Southwestern U.S. was still part of Mexico.

In the Northwest, some Euroamerican people in the state of Montana have the same prejudice, and a few years ago pushed to have "English only" laws passed, especially but not exclusively to stop the use of Spanish on state documents, judicial proceedings, etc. This was ironic for three reasons, in particular (not counting the enthusiastic reception and enjoyment of Mexican food and Indian fry bread). First, as stated in data available from the state's Office of Public Instruction (OPI), Montana high schools were required to offer one unit of a second language in 1988, which was raised to two units in 1991; middle schools and junior high schools were required to offer a half-unit beginning in 1994; for all levels, the terms "foreign language" and "second language" are now replaced with "world languages" which include "American Indian Languages." Education and business interests have recognized the commercial and cultural benefits of people speaking more than one language. Second, the name *Montana* is a corruption of the Spanish word *montaña*, "mountain" (which is noted in a description of the state seal on the secretary of state's web page), and a mountain range is in the Great Seal of the State of Montana (as is the Great Falls of the Missouri River, the longest river in the U.S., which originates in Montana; the Great Falls, named by Lewis and Clark on their Voyage of Discovery, later disappeared underwater because of dam construction); and, the seal has, in Spanish, "oro y plata" ("gold and silver") and a drawing of miner's tools and a farmer's plow. The linguistic irony of Spanish words in the state name and seal while "English only" efforts were underway was noted: did advocates of ethnic bias want to change the state seal? Third, the real "native languages" are native Indian languages (there are seven Indian reservations in Montana), spoken by the only native Montanans; therefore, the only "native speakers" are the native peoples who preceded Europeans and Euroamericans.

In recent decades native Indian peoples have sought to save their slowly disappearing traditional languages through recorded interviews with elders, from which oral histories, dictionaries, and pronunciation are compiled; this effort is complemented by classes taught by elders and others to help the youth and even adults learn their traditional languages and recover aspects of their culture, since language is a conveyor of culture. The Constitution of the State of Montana, revised in 1972, requires support for native cultures not only on their reservations, but in public school curricula across the state; Article X, Education and Public Lands, Section 1 (2) declares that "the state recognizes the distinct and unique cultural heritage of the American Indians and is committed in its educational goals to the preservation of their cultural integrity."

So, the recently arrived Euroamerican descendants of people who had used Discovery to seize lands and goods from Indians now seek to force "English only" on the real Native Americans and native Montanans, part of an ongoing U.S. effort to promote cultural genocide. This has been the case not only with the real "natives." A similar impact is being made to impose "English only" on latinas/os in particular (but also on Asians), among current and more recently arrived immigrants, thereby promoting cultural genocide. *Xenophobia* runs rampant among some U.S. citizens. A "fear of the Other" has become translated into cultural repression and economic and political oppression of those considered "aliens"—by those whose ancestors are the real "aliens" and foreigners, relative to the thousands of years native peoples lived in the U.S., and the hundreds of years Mexicans and

Hispanics have been inhabitants of these lands (St. Augustine, Florida, was established by Spain in 1565—Spain ruled Florida and the southeastern area of what is now the U.S. for two centuries—and is known as the "Oldest [European-founded] City in the U.S."; and, of course, an explorer commissioned by Spain, Christopher Columbus, "discovered" the "New World"). Rejection of an "other" solely because they are "Other" is, of course, contrary to the teaching of Jesus whom many of these proponents cite as the one they claim to follow as "Christians." As noted previously, the religions of Christianity, Judaism, and Islam, and native Indian spirituality teach *xenophilia*, love of the "other," the "alien." The Hebrew Bible reminds Jews that they were once "aliens" in Egypt and, having suffered oppression there, they should not act similarly toward the "alien" in Israel. A variation of a biblical teaching cited earlier would serve us well when considering past history in light of the here and now: "Do not do unto others what you would that others had not done unto you."

So, What's the Bottom Line on Aliens?

We've seen that alien plants, animals, people, and extraterrestrials might be either accepted into or prevented from entering our lives and communities. On first encounter, they are "alien": "the Other," "the stranger," "the foreigner," someone or some biotic entity or even an abiotic place that is unfamiliar and "strange" at first. Sometimes the unknown can catalyze fear, resistance, and even violent confrontation. This might be a deeply embedded caution and fear that arose early on in our evolution as a species, carried from our previous biotic existence when basic survival was imperative and at times nebulous and perilous, and instinct and experience instilled them in us; a similar consciousness was subsequently carried forward consciously or subconsciously into our present state of biological and cultural evolution. (This was not the case of all peoples and cultures: remember the hospitality of many natives of Turtle Island when European explorers arrived.) Some people have overcome the primal fear through individual or social changes in consciousness and conduct and, though realistically cautious in encounters with previously unknown—"alien"—people and peoples when these others appear outwardly hostile and acquisitive, have a core welcoming and curious attitude toward new arrivals on the scene. We've seen how we've integrated aesthetically pleasing (in gardens, landscapes, or as found in pristine places), helpful, and even life-giving plants (as food and medicine), have domesticated previously free ("wild") animals to provide food or clothing (cows for milk, chicken for eggs, sheep for wool; all three animals as food, for some of us), and have accepted and integrated—not, at times, without conflict—people not of our own ethnic group, language, culture, economic status, religion, or sexual orientation. "We" and "they" have become "us," so to speak.

Perhaps, similarly, with ETI. Presently, we do not know if the ETI that comes into sustained Contact with us will be benevolent or malevolent; recognize and accept the intrinsic value of Earth and Earth's biota, and be interested in our species' and our planet's wellbeing, or regard Earth's natural goods, territories, and biota to have only instrumental value for ETI, to be taken by whatever means necessary, as they put into practice their own theory, an ETI Discovery Doctrine. Indeed, as Stephen Hawking fears, just as Contact between native peoples of Turtle Island and alien Europeans "did not turn out very well for the Native Americans," so too Contact between human beings and invading ETI would "not turn out very well for humankind." At this point in human history, when we might well be billions of Earth years younger than ETI explorers, we do not know the outcome of Contact. To date, ETI has not used weapons belligerently or defensively when fired upon by TI military personnel representing diverse nations in distinct places (Leslie Kean, in *UFOs: Generals, Pilots, and Government Officials Go on the Record*, describes aggressive actions by air force pilots from Iran and Peru, among other countries, who fired missiles and machine guns at alien spacecraft). This might indicate that ETI is not aggressive or acquisitive and is curious about our current state of development as a species. Perhaps, too, ETI is interested in our wellbeing and hopeful that we can eliminate potentially genocidal wars among ourselves and our ethnic and economic oppression of the "least of these" in our nations and internationally.

We should remain conscious of the fact that when we venture off our planet, in our exoEarth voyages and arrivals on other worlds we are to the worlds and whatever life we might encounter "extraterrestrials": we are the arriving "aliens" entering an alien (to us) land, and encountering alien (to us) biota, both single-celled and simple or highly complex and intelligent, to the extent of whatever degree of consciousness they have of our alien presence. They might regard human aliens as congenial and collaborative or conflictive and invasive.

ENCOUNTERING ETI

When TI becomes ETI on another world, as happens in fiction in *Avatar* when humans arrive on Pandora and Contact the native Navi, and in light, too, of the fears that some human beings have about what will happen when ETI arrives on Earth, another variation of the paraphrased biblical verse used previously is very fitting regarding TI-ETI Contact, wherever it occurs: "Do not do unto others what you would not have others do unto you."

We'll discuss extraterrestrial aliens in greater depth in chapters to follow.

2

EXTRATERRESTRIALS: THEIRS AND OURS

Are We the Others We Don't Want Others to Be?

PEOPLE GENERALLY ARE ACCUSTOMED to thinking of an "extraterrestrial" as someone who is exoEarth, from a place beyond planet Earth, an "alien from outer space." The meaning of extraterrestrial can include that and more. When we're in space, we'll be "extraterrestrials": we will be "off Earth" in relation to our own home planet and other evolved members of our human species. On Earth, it could designate anyone who comes from a place elsewhere on Earth, who is exoterritorial in terms of our particular territory (our place on *terra*, Earth), someone who is beyond a particular group's or nation's resident space. It would also be how we humans would appear to extraterrestrial life (ETL) and extraterrestrial intelligent life (ETI) when we arrive on their "*terra*," their particular planet or region of their planet. We would be in ETL's perception a previously unknown intruder into their habitat, and in ETI's view the "extraterrestrials," the "aliens" arriving on and perhaps intending to extensively invade their space. Intelligent beings on distant planets might, in fact, regard us the way that Indians and other indigenous peoples perceived Europeans who initially explored and then colonized Indian territory—and then took both Indians' territory and their natural goods.

In terms of terrestrial-extraterrestrial designation, then, if we come into Contact on our Earth home *they* are *our* extraterresrials; if our Contact takes place on their home place in space *we* are *their* "extraterrestrials."

While *terra* means "Earth," then, it might also mean territory (terratory → territory) or land (earth). A convenient convention to be used in these pages is that when distant planets or other celestial bodies become part of human consciousness, and are seen as places in which to settle, "territory" will designate such places: the land and the natural goods ("resources") that are present in place. Intelligent beings on other worlds who discover alien humans exploring and landing on their territory will regard the humans as "extraterrestrials," that is, "extra" or externally originating relative to intelligent inhabitants' places: humans will be ETI, and resident intelligent beings will be TI (where "terra," in such instances, means the home world of any life).

Humans on distant planets might, if we do not change our current consciousness and conduct, act just as Europeans did in the fifteenth to seventeenth centuries in newly found regions, including in what they called a "New World." We might explore and interact with others influenced by our conscious or unconscious Discovery Doctrine. Europeans did; it was appropriated subsequently by Euroamericans after the Revolutionary War. With Discovery in our mind on distant worlds, we might then attempt to use superior technology and weaponry, should we have that, to overcome through coercive violence, to the extent that we deem it necessary, any native objection or resistance to human imperial ambitions regarding territory and its natural goods.

Discovery Doctrine: Will Humankind Do in the Heavens as We Have Done on Earth?

Will history repeat itself in our space explorations, such that Discovery runs rampant and self-justifies our conduct as we decimate resident alien peoples, devastate their natural and cultural environments, and deprive them of their needed sustenance goods? Will Discovery on Earth be Discovery redux in the heavens, such that humans' will is done "in the heavens as it is on Earth"? Or, conversely, will we have undergone a political, economic, and spiritual *metanoia*, a conversion of mind and heart, prior to our arrival elsewhere in the cosmos such that our consciousness and conduct will have been changed for the better?

Put another way: will we or will we not be the type of "alien" we and exoEarth residents fear will arrive on our respective home planets, to our great detriment? Will we be there, on others' planets, the type of aliens that we don't want others to be here, on our planet?

Science Fact

The existence or not of intelligent extraterrestrial beings continues to be a matter of conjecture and debate for most people on Earth. Those who understand that they have seen ETI, or evidence of ETI existence, especially as represented by space vehicles whose technology far surpasses what humankind has developed to date, regard themselves to be beyond "belief" and in the realm of "knowledge." They do not *believe* that UFOs and, by extension, that ETI exists: they *know* that extraterrestrial spacecraft and intelligent life exist. (The next chapter explores and elaborates this in depth.) They assert that what they know is scientific fact. Those lacking a similar experience are often skeptical about, or even deride, UFO/ETI reports and those who present them. It is scientific fact that objects have been viewed, tracked on radar, and filmed traveling at phenomenal speeds and exercising intricate and dramatic maneuvers not previously seen or even theorized in the past. It is still scientific speculation, however, about whether or not these objects are inhabited by intelligent alien explorers or controlled from afar by alien intelligent explorers, perhaps by computer control or via robots; it will become scientific certainty if scientists obtain hard evidence of some form that confirms UFO and ETI existence and presence. Some scientists apparently have confirmation already from debris and the data it provided if, as has been speculated for decades—including by foreign governments' officials—the U.S. has sequestered and is analyzing debris from a 1947 Roswell, New Mexico crash of two alien "flying disks" as initially reported in a press release by Colonel William Blanchard, base commander of the Roswell Army Air Field. Blanchard had sent base security personnel to the reported crash site on the Foster Ranch, and issued the statement after they returned with items they had found (see chapter 3).

UFOs: Fact and Fiction about Close Encounters of the Alien Kind

Over many centuries on Earth, people have come to accept the existence of space explorers. Their decisions to do so were based on a variety of factors.

Some had a visual experience; others claimed to have had direct Contact of some sort; others accepted the word of a friend, family member, or other credible source who stated that they experienced Contact or knew of another's Contact experience; yet others, in recent centuries, read a news or tabloid story, or saw a television documentary on the topic. The extent to which people in the U.S. are open to the idea of Contact is evident in the immense popularity of the 2009 3-D film *Avatar*, earlier films such as *E.T. the Extra-Terrestrial* and *Close Encounters of the Third Kind*, and the various *Star Wars* films and episodes of the *Star Trek* television series in its various iterations. Most dramatically, however, people's acceptance of the possibility of extraterrestrial alien visitation, and even of hostile extraterrestrial aliens' invasion, was evidenced by the anxiety and panic that ensued when Orson Welles directed and narrated an adaptation of the H. G. Wells novel *The War of the Worlds* for the CBS radio network (described in the next chapter). The current rash of violent films and television programs depicting malevolent aliens destroying Earth cities and massacring human populations—and, of course, to which a surviving band of heroic, paramilitary human, mutant human, or humanoid individuals and groups respond with ongoing guerrilla tactics—seems intent on preying upon humans' worst fears. The minority of aliens in such films who are benevolent and helpful is a token gesture to the idea that perhaps not all—and perhaps not any—aliens are angry, aggressive, and acquisitive.

In the latter part of the twentieth century—unfortunately, in the perspective of actual witnesses—the credibility of UFO events diminished considerably. The causes included multiple unverifiable stories about alien abduction; individuals' claims that they were designated by superior ETI beings to "channel" the ideas of advanced alien sages who represent a higher level of ETI and are speaking to humankind through human channels, from near or distant space, in an attempt to save people from self-destruction and, concurrently, save Earth from being destroyed by humankind; the formation of religion cults, such as Heaven's Gate, oriented toward supposed ETI teachings; U.S. government efforts to suppress news stories about and scientific inquiries into UFO encounters and, as an additional measure, to use public ridicule or the fear thereof to coerce even credible witnesses into silence—even when these witnesses know that they had seen a UFO; and denial of the existence of UFOs and ETI in institutions of higher education, and university-related (and often U.S. government- and U.S. military-funded) research bodies, accompanied by ridicule of faculty (no matter how senior in rank) and denial of tenure to faculty or an academic appointment to potential faculty, who merely mention in casual conversation that they are curious about UFOs and ETI.

Pause to ponder the latter paragraph: science, which is supposed to have some "objectivity" in its research methods and creativity in exploring new avenues and topics for research, is forbidden to do so in the case of unidentified flying objects (UFOs), which have more recently been termed unidentified aerial phenomena/on (UAP). (The latter, more generic term lacks the "baggage" of popular use of UFO to mean automatically a space vehicle occupied by extraterrestrial intelligent beings.) Should not scientists be encouraged and funded to explore with careful, expert-based objectivity and the most technologically advanced instruments possible, any UFO/UAP narratives? Such dedicated and serious scientific investigations would help confirm which events are just natural occurrences in the heavens (meteors, lunar passage, Venus, etc.) or human artifacts (government or private company satellites), and which are true anomalies that do not fit into those categories—and, if this is so, why. What characteristics do they possess that place them beyond what is known in current human knowledge, and beyond current human experiences and technological creativity? There are numerous credible witness reports, astrophysicist J. Allen Hynek's accounts of his findings in more than two decades as the USAF civilian scientific consultant (see below; for much more detail, see *Cosmic Commons*), official government reports recently released (often grudgingly), and still-occurring events, as reported around the globe. Some UAP activities are being recorded by armed forces members in several nations, who use civilian and military radar and the photos or other data given to them by air force personnel to try to ascertain what has been seen and reported and evaluate it. An ordinary, thinking citizen feels compelled to ask "Why?" as they wonder what rational reason(s) there might be not to explore and subject to scrutiny what must be, for scientists in particular, an exciting array of data. The "transparency" people want in their government's operations should include transparency vis-à-vis UFO/UAP/ETI. Transparency on these matters is operative today to a greater extent than previously in most nations around the world. Transparency in UFO/UAP/ETI matters is *not* operative in the U.S., or in regard to U.S. military personnel reports of encounters on foreign soil.

The Extraordinary (but not Extraterrestrial) Josef Allen Hynek, UFO Supergenius

Okay, let's be clear right out of the chute that we're not talking about an extraterrestrial from Mars or Alpha Centauri who came to Earth as a benevolent being, someone who (1) brilliantly pointed out our human flaws and failings and showed us a different way to think and act, (2) brought us technologies that enabled humankind to take a quantum leap in research and development and better our civilization's commercial and industrial expertise and accomplishments, (3) was all of the stereotypes of the ideal television emergency room doctor and of medical lab researchers working on eliminating life-threatening diseases, or (4) all of the preceding rolled into one, for the betterment of human health and overall human societal and personal wellbeing. The UFO supergenius cited in the heading was born right here on Earth. People around the world have benefitted, in ways unknown to most, from his scientific intellect, curiosity (fortunately, he wasn't born into a political culture where a "know-nothing" cadre of content controllers banned curiosity and analytical thinking from school classrooms and government legislative sessions), and inventive creativity, both technologically and theoretically. We're speaking in this context, of course, about Dr. J. Allen Hynek (1910–86), who came to be the leading UFO researcher in the world. Hynek developed the three categories of "close encounters" that are still used today, decades after their first formulation. A very fascinating, thought-provoking, and credible person, Hynek was an accomplished astrophysicist and astronomer who began work in the field of UFO studies as a civilian scientific consultant to the U.S. Air Force. He investigated perceived UFO phenomena and claims that people presented regarding Contact with ETI, and through more than two decades served as the official U.S. government scientific expert who denied the veracity of any reports about UFOs and ET. Over time, however, after probing further and reflecting extensively on the cases and evidence that he had analyzed and, at times, ridiculed, he altered his longstanding position on UFOs.

No intellectual "lightweight" or academic slouch, Hynek received his BS (1931) and his PhD (1935) in astrophysics from the University of Chicago, where he did his core doctoral studies at the Yerkes Observatory. Over the course of his career he was Professor, Department of Physics and Astronomy, Ohio State University; research scientist on satellite tracking (he developed the system used by the U.S. to track satellites, which was copied by the U.S.S.R.), Smithsonian Astrophysical Observatory; and Professor and Chair of the Astronomy Department, Northwestern University. He authored numerous scientific books and articles in astrophysics (and, later on, regarding UFOs) and wrote the astronomy column for *Science Digest*.

As the scientific consultant and principal astronomy expert for U.S. Air Force projects focused on UFOs (1947–69), Hynek was supposed to determine whether or not people's perceptions of UFOs were misidentifications of a natural phenomenon (such as an unusually bright meteor), human artifacts in the atmosphere (satellites, space debris), or visual sightings of actual UFOs under intelligent control. The undisclosed actual functions of his office and of air force projects (as acknowledged by Hynek after he resigned from his position) were to collect and store data on UFOs; deny that UFOs exist, or that they had appeared in U.S. skies; dissuade people from thinking that they had seen a UFO—even when (and especially when) the air force and the U.S. government knew that a UFO event had occurred at the time and place witnesses stated that it had; fabricate false explanations for UFOs when they were known by the air force to have been actual aerial anomalies or perhaps space-originated vehicles; and use ridicule in press conferences and news releases, among other places, so that people became concerned about possible professional and personal adverse consequences if they acknowledged they had viewed a UFO even from afar. Consequently, highly respected university professors feared to note their scientific interest in UFO phenomena, let alone their personal sighting of one, for fear that family, friends, and colleagues would ridicule them.

Now, imagine yourself as a highly educated scientist who demanded meticulous evidence to establish working scientific hypotheses and theories, and eventually to formulate and present scientific "facts" that would be accepted throughout the scientific world and ultimately by the general public. As you ponder possibly exploring this topic, perhaps you dream of a Nobel Prize should you have a research breakthrough. As you continue to think, intruding into your intellectual landscape and interfering with whatever settled scientific facts and theories you have and whatever core metaphysical beliefs you might have, claims of sightings of otherworldly, technologically superior spacecraft thrust themselves into your musings and consciousness and press against your existing stable and secure views. How would you react? Probably much the same as Hynek did.

Hynek thought that the idea of UFOs and ETI was ridiculous, and that people who believed in them must have mistakenly thought that some meteorological phenomenon, lunar or planetary body motion or reflection, or other naturally explainable existent or event was a UFO; or, that they were mentally unbalanced in some way. He scoffed at such "erroneous" thinking and welcomed the USAF's invitation to be their official scientific consultant, with the role of "debunker" (his term and the USAF term) of UFO narratives. He relished his role enthusiastically, in the interests of science and the public interest. For all his enthusiasm, once he became aware of the U.S. government's use of coercion and ridicule to suppress UFO stories Hynek did not agree with such tactics, believing that science could readily perform the same task. He began to urge ever more forcefully that the USAF and the government should fund and support objective scientific research into the phenomena: he was sure that science would irrefutably establish that UFOs, understood in public thinking as intelligently operated and occupied spacecraft, did not exist.

However, the unexpected entered Hynek's horizon, in the form of . . . scientific fact. Even after his rigorous scientific investigations into UFO reports, and his dismissal of most—95 percent—as natural phenomena, he realized that while 95 percent is an overwhelming percentage, 5 percent of several thousand reports remained and lingered, with no scientific fact or theory available to explain them. He decided to undertake a secret survey of fellow astronomers around the U.S. who were highly regarded professionally, and would easily have been able to distinguish among natural phenomena of whatever sort, Earth-launched satellites, and anomalies that could not be described accurately and dismissed because they violated known laws and theories of physics and known, envisioned, or projected human technologies. When he sent out the secret survey selectively and individually, Hynek assured each scientist that their name or professional work and affiliation would not be revealed, and noted that there was no request on the survey for that information. Some were possibly still fearful of consequences of participation, should their personal or professional data be discovered in some way, and did not return the secret survey; others were definitely fearful, and emphasized that neither their name nor occupation should be used in disseminating survey results, even confidentially among survey participants or other members of the scientific community. Hynek learned from survey respondents that 11 percent of the astronomers had observed anomalous objects and events—things that could not be scientifically explained. He became increasingly dissatisfied with the air force's refusal to allow objective thinking and scientific methods to investigate UFO reports, resigned from his position, and embarked on his own science-based UFO studies.

Several years after Allen Hynek's resignation the USAF, pressed more intensely by the public and the press for information about UFO incidents, declared unilaterally that there was no such thing as a UFO, and that it would not investigate the matter further or gather any data about reported events. The press, for decades, followed the USAF's cue and did not report on any UFO events except, on occasion, bemusedly. Author Leslie Kean observes in her book *UFOs: Generals, Pilots, and Government Officials Go On the Record* that, based on her analysis of foreign government documents, her interviews with foreign officials, and stories in the international press, there is widespread belief at the highest levels of foreign governments that the USAF, the CIA, and other, unknown U.S. government secret agencies and secretive individuals have been reviewing, filing, and storing information on UFO events around the globe. In some cases, agents of the U.S. government apparently coerced foreign officials and citizens in order to suppress UFO research, reports, and release or elaboration of news about UFO sightings or close encounters.

Hynek discusses his UFO efforts extensively in two books, *The UFO Experience: A Scientific Inquiry* (1972) and *The Hynek UFO Report* (1977). In *UFO Experience*, he defines "UFO" as simply "the reported perception of an object or light seen in the sky or upon the land the appearance, trajectory, and general dynamic and luminescent behavior of which do not suggest a logical, conventional explanation and which is not only mystifying to the original percipients but remains unidentified after close scrutiny of all available evidence by persons who are technically capable of making a common-sense identification, if one is possible." The short form of Hynek's UFO definition might be this: after rigorous scientific scrutiny of an incident and careful analysis of data collected, any unusual object in the sky that remains unexplained can be acknowledged to be a UFO.

In the same book, Hynek elaborates six categories of distinct events in which credible witnesses report UFO experiences, and how near or far the witnesses were from the objects they sighted. Distant objects were classified as nocturnal lights, daylight discs, or radar-visual events. Objects near the observer(s) were placed in the category of "close encounters." The latter classification gained prominence among the general public when Steven Spielberg used one of the categories as the title for his film *Close Encounters of the Third Kind* (1977), for which Hynek was a consultant, and in which he had a brief cameo at the end.

Hynek's "close encounter" categories were "close encounters of the first kind" (CE-I), in which the observer sees an object less than five hundred feet away (for you sports fans, that's less than the length of two football fields); "close encounters of the second kind" (CE-II), in which physical evidence of a UFO occurrence is found at the site where the UFO was thought to have landed; and "close encounters of the third kind" (CE-III), in which humans come into direct Contact with extraterrestrials (as individuals or collectively, as portrayed in Spielberg's movie).

Close Encounters of the Second Kind particularly interest scientists who can, in a sense, bring the UFO into their research facility. Evidence gathered on-site, such as burnt grasses, samples of disturbed soil, etc., can be tested with a view toward determining what caused the burn, what pressures were necessary to produce the imprints on the ground, and what chemical changes could be discerned between on-site soil samples taken from the affected area, compared to control samples gathered from the surrounding area. Hynek identified more than eight hundred cases where a visual report of a UFO in the sky correlated with physical evidence of a UFO presence on the ground.

Close Encounters of the Third Kind, when reported by credible witnesses, are the most mind-boggling: highly intelligent beings who are far more advanced technologically than humans and are not native to Earth are making direct Contact with us. Hynek noted that credible witnesses' CE-III reports disturbed him most psychologically because of their implications for humans' self-identity and their sense of their place in the universe. People resist learning about such threats to self and species much as their ancestors reacted to the idea that Earth was part of a solar system, centered on a star, and not the center of the universe; past and present assertions about the nature and importance of Earth's fossils; and theories that meteors were space rocks. Hynek studied, in depth, close encounters experienced in New York State's Hudson River Valley and its surrounding region (as will be discussed in chapter 3). Readers who wonder what Hynek looked like in real life can see him briefly toward the end of the film *Close Encounters of the Third Kind* among the scientists and government officials who are slowly walking down the runway to meet ETI representatives from the landed UFO. He is dead center in the screen, walking directly toward the camera through a crowd of scientists. He is sixtyish, with a mustache, goatee, and brown-rimmed glasses, dressed in a grey suit, white shirt, and dark tie; he stands out from the lab-coated scientists around him. He stops walking, strokes his beard with his left hand, and then puts a pipe in his mouth with his right hand. This is his brief moment of screen "stardom"; his writings provided resource material for the film. The film's and his books' exploration of potential Contact with intelligent beings from the stars, and actual UFO events reported by the public and described in the press (usually, only locally), stimulated ongoing, on-the-ground public fascination with UFOs and ETI—and concern about implications of their presence in the cosmos and on or near Earth.

Thoughtful Consideration of Encounters

The depth and breadth of the scientific body of work produced by J. Allen Hynek stimulated significant intellectual exploration, usually clandestinely for professional and personal reasons stated above. His efforts provided an excellent base for a focused consideration of plausible accounts of extraterrestrial visits. As Hynek notes, to come to ascertain the meaning of unidentified aerial phenomena and the intelligent beings who operate them requires a twofold process—understanding UFOs and understanding ETI. His proposal could be explored through three questions related to his and his successors' work: What explanations might be suggested for UFO phenomena capabilities and origins when UFOs are perceived as nocturnal lights or daytime discs (and other shapes), in radar-visual sightings, and in the categories of "close encounters"? What intelligence controls them—is it ETI? Have there been scientifically confirmed verifiable visits by ETI? The search for responses to these questions should excite scientists in multiple disciplines of the physical sciences, social sciences, and humanities. When the research takes place and bears fruit, then even in the face of ridicule and other adverse reactions, humans will have made a giant leap forward not only in understanding the cosmos, but in finding their niche, voice, and meaning in it.

Credible Witnesses: To Believe or Not to Believe

Credible witness reports of UFO experiences help promote—and provoke—consideration and conversation about UFOs and ETI. Hynek reported that there had been thousands of cases throughout the world by 1972, including in the U.S.—and today we are more than forty years beyond the report period he cites. Reports continue to be made: while this occurs because of much greater openness by foreign governments and more coverage in the foreign press, it has been assisted today, too, because even major U.S. news networks, the most credible newspapers (beyond sensationalist newspapers), and television programs on stations such as PBS and the Discovery Channel—as well as science fiction series and stories on other channels—cover and release nonfiction witness reports (and occasional video or still photos) about such incidents, or fiction movies based on them.

Science Fiction

While research into possible science fact strives to sustain a focused, accurate presentation of available data, and objective analyses and assessments of UFO and ETI narratives, science fiction roams the realms of diverse imagined scenarios regarding possibilities of terrestrial-extraterrestrial intelligent life Contact. Carl Sagan, NASA consultant, Professor of Astronomy and Space Sciences at Cornell University, award-winning scientist, and PBS TV personality, bridged the realms of science fact and science fiction. He used objective, science-based narratives in his television series, *Cosmos*, his nonfiction writing such as *Dragons of Eden: Speculations on the Evolution of Human Intelligence* (for which he received a Pulitzer Prize), and his novel, *Contact*, which was made into a popular movie with the same title, starring Jodi Foster. Our science fiction considerations will be limited to just a handful of examples available from a vast literature that goes back at least into the nineteenth century. We'll discuss novels and movies.

The Martian Chronicles

Pulitzer Prize recipient Ray Bradbury (1920–2012), a writer of fantasy and poetry, describes in *The Martian Chronicles* (parts of which were published first as short stories in periodicals) humankind's exploration of and migration to what seemed to be the nearest planet in our solar system that might have (or once had) life. In his novel, as in numerous other works of fiction related to the "red planet," Martian life evolved to include intelligent life, ETI. (Page references in parentheses are to the Harper Perennial edition, 2011.)

During a period of severe political conflict on Earth, when humankind seems to be heading inevitably toward nuclear war, exclusively U.S. representatives of humanity are sent on exploratory expeditions to Mars. The first one occurs in 2030 and, as might be expected, is a military expedition. Proud of their accomplishment in being the first humans on Mars, the explorers are disconcerted when resident Martians are not impressed. Mars inhabitants think that the explorers and their rocket are figments of the imagination of the Earth commander, who is assessed to be an insane Martian able to project the supposed humans' existence; this is in turn reified by the telepathic capabilities of the Martians who believe they are seeing events through the eyes and under the coercive influence of their insane citizen. Consequently, the human explorers are killed because the commanding officer's mental illness seems contagious to the Martians when they begin to think that the mentally projected invaders do, in fact, look real. The Martians themselves lacked space exploration technology, so interplanetary travel was not a part of their thinking, but they had evolved highly developed mental abilities, including telepathic reading of others' minds.

As the stories unfold and describe subsequent Mars expeditions, humans continue to arrive solely from the U.S., with ever more sophisticated technology and weaponry; they perpetrate genocide against the Martians in order to seize their territories and natural goods (this is an unstated expression of Euroamericans' use of Discovery Doctrine to devastate native peoples); eventually the U.S. sends immigrants by the thousands to colonize Mars and terraform it (alter it to be Earthlike) in order to satisfy human needs and wants. Mars must look like Earth, in their view, the more easily to be inviting to and culturally accepted by U.S. citizens journeying great distances from Earth. Over time, global wars erupt on Earth, and all the immigrants-now-residents are forced to return to Earth by their government. Only two remain, a man and a woman, but they do not become a new, Martian "Adam" and "Eve," due to their personality conflicts, and live and die alone.

Bradbury's distinct and distinctive Martian short stories were written in the aftermath of World War II. Subsequently, they were linked together in novel format, and provide an insightful, running commentary on socioecological issues of the period and their related conflicts, and how they would be projected onto distant settlements. It remains a realistic and pessimistic assessment today of what could transpire, given current thinking and desires. It contradicts Stephen Hawking's optimism about the prospects for humankind's representatives who will be sent into space to colonize Mars. Hawking's pioneers, as described earlier, are to be whisked away by a *deus ex machina* spaceship, a contemporary Noah's Ark and lifeship whose construction and launch he advocates in order to save a segment of the species from the social devastation and ecological destruction that humans developed and unleashed on Earth. The ideas, incidents, and ideals in Bradbury's narration, by contrast, provide deeper insights regarding Contact and its aftermath, which are especially relevant for contemporary, twenty-first-century consideration when humankind is almost at the dawn of its Star Age, the successor to its Space Age.

Bradbury's concerns and implied or overt social critiques and religious insights are multiple (however unintentional or intentional some might be). They include the following:

- *Ethno-cultural imperialism* (a form of the Discovery Doctrine, although Bradbury does not name it): The captain on the second expedition, when communicating with a Martian woman who tells him to whom he should speak in town to advise him, responds, "We don't want to know anything. . . . We already *know* it" (25).

- *Expectation of a devastating global atomic war*: When the third expedition, primarily a military one, is about to leave Earth, a man who tries to join the colonists-to-be is constrained, and yells, "Don't leave me here on this terrible world, I've got to get away; there's going to be an atom war! Don't leave me on Earth!" (42).

- *Social scientific and religious anthropomorphic projection*: When the expedition lands, they find, unexpectedly, that Martian cities are exactly like Earth cities, and the people appear human, some of them long dead relatives of crew members: the crew discusses whether they are "on the threshold of the greatest psychological and metaphysical discovery of our age!" or if "we're looking upon a phenomenon that, for the first time, would absolutely prove the existence of God" (47). However, the reality is that the Martians used their telepathic powers to project a virtual reality based upon the explorers' memories in order to lull them into a relaxed state of mind—and then kill them.

- *Interworld pathogenic transfer*: In the next expedition, humans find that a harmful Earth microorganism had been inflicted on Mars biota beginning when the previous expedition arrived; all Martians are

Extraterrestrials: Theirs and Ours

dead, killed by chicken pox against which they had no immunity (69), an impact reminiscent of the smallpox and hepatitis brought to the Americas by Discovery-era European explorers and colonizers, and wrought upon native inhabitants.

- *Ideological and ethnocentric detachment from place and ethnocentric cultural imperialism*: When the soldiers have an exuberant alcohol-assisted celebration the archaeologist with them objects, complaining to his commander that his companions lack respect for the deceased Martians and their civilization, and for the "spirit of things as they were used" which still remains; he elaborates (in words alluding, perhaps unintentionally, to the material and cultural genocide unleashed against Indians in the Americas in the aftermath of Discovery): "All the things which had uses. All the mountains which had names. And we'll never be able to use them without feeling uncomfortable. And somehow the mountains will never sound right to us; we'll give them new names. . . . The names we'll give to the canals and the mountains and cities will fall like so much water on the back of a mallard. No matter how we touch Mars, we'll never touch it. And then we'll get mad at it, and you know what we'll do? We'll rip it up, rip the skin off, and change it to fit ourselves. . . . We Earth Men have a talent for ruining big, beautiful things" (74). To which the officer replies, "One day Earth will be as Mars is today. . . . It's an object lesson in civilizations. We'll learn from Mars" (75).

Bradbury explores other scientific, geopolitical, anthropological, and sociological issues, such as:

- *The possibility of transpermic transfer*: The humans pondered, "How had they built this city to last the ages through, and had they ever come to Earth? Were they ancestors of Earth Men ten thousand years removed?" (77). (The Greek philosopher Anaxagoras wondered if cosmic sperm migrations seeded worlds, and termed this *panspermia*; today, such a possibility is labeled the "panspermia hypothesis" by scientists, who speculate about the possibility that life has traveled through space as microbes, and seeded planets in the solar system, the galaxy, and even the extended universe. The recent sci-fi film *Prometheus* has a variation on this theory, especially evident in the opening spacecraft and waterfall scenes.)

- *Cross-cultural conversion*: Spender, the expedition archaeologist, goes off by himself, studies Martian writings, and, seemingly telepathically or perhaps actually transformed, returns and kills those of the soldiers he sees; he then flees to the relative, albeit temporary, safety of the mountains. Later, he meets the commander there and adds a sociocultural commentary—"anything that's strange is no good to the average American. If it doesn't have Chicago plumbing, it's nonsense"—and a caution about the *nationalistic, military intent of space exploration*: "You heard the congressional speeches before we left. If things work out they hope to establish three atomic research and atom bomb depots on Mars. That means Mars is finished" (88). Spender is killed shortly thereafter. His observation was borne out in 1959 when the U.S. Army secretly designed and requested funding for a moon military base, Project Horizon, to establish U.S. claims to the moon and to defend Earth against ETI.

- *Immigrant pioneers' hope-filled migration*: Settlers are attracted to Mars by government promises of work on Mars (99), similar to what was done in the U.S. to motivate European immigrants to move westward (despite existing native peoples' communities, who had signed treaties that guaranteed them that they would retain their accustomed places and territorial goods).

- *A greening of a planet*: Not of Earth from which emigrants would come—Earth was believed to be on the verge of atomic war—but of Mars. On Earth, people did not attempt to *change* humans' Earth consciousness and conduct but, like Hawking's projected emigrants, sought to *escape* from humans' destructive impacts on their home planet—*terraforming* Mars to make it like Earth (100–106).

- *The urbanization of landscapes*: "From the rockets ran men with hammers in their hands to beat the strange world into a shape that was familiar to the eye, to bludgeon away all the strangeness, their mouths fringed with nails so they resembled steel-toothed carnivores" (107).

The colonizers in their hubris exemplify the following:

- *Nationalistic technological superiority and exclusivity*: Only people from the U.S. are permitted to travel to and colonize Mars to find a better life because the "rest of the world was buried in war or the thoughts of war" (119).

Extraterrestrials: Theirs and Ours

- *Missionary cultural and religious arrogance and exclusivity that denigrates and denies others' faith traditions*: An Episcopal bishop addressing his priests as they prepare to depart for Mars tells them, "I know you will go with God, to prepare the Martians for the reception of His Truth" (123).

- *Anthropomorphic religio-moral projection*: On landing, in their prayer giving thanks, the missionaries observe that "there will be new sins," walk to the city mindful that they are in the "territory of new sin" (124), and ask the mayor about the physical characteristics of Martians so that they might properly reproduce them in the stained glass windows of the church to be built in those parts of their ancestral territory where Martians still dwell, primarily out of sight. The missionaries learn about *unexpected new forms of intelligent beings*: the mayor tells the priests that there is one species recently encountered which has the form of "blue spheres of light" (125); subsequently, when the blue beings appear, the priests argue: "God's work!" versus "the devil's!" (128).

- *Refusal to acknowledge natives' benevolence*: When the blue spheres save the priests' leaders (appropriately surnamed Father Peregrine and Father Stone) from being crushed by an earthquake, one states that the spheres saved them, since they were suddenly two hundred feet away from the avalanche, while the other asserts the two escaped by running away (130).

- *Theological projection onto the other*: "They saved us. They *think*. They had a choice; let us live or die. That proves free will!" (131). (Note the irony and role reversal: the missionaries went to Mars to "save the souls" of Martians, but the Martians materially saved them—just as Indians had saved starving Pilgrims.)

- *Nonanthropomorphic theological realization*: One missionary states, "No Adam and Eve on Mars. No Original Sin. Maybe the Martians live in a state of God's grace," to which his companion, frozen in *theological anthropomorphism*, replies, "Yes, Father Stone, but the Martians killed some of our settlers. That's sinful. There must have been an Original Sin and a Martian Adam and Eve. We'll find them. Men are men, unfortunately, no matter what their shape, and inclined to sin" (131–32).

- *Temptation, sin, and salvation*: In a direct contradiction to Jesus' decision when tempted by Satan in the desert to jump from the temple pinnacle in order to be saved by angels, Father Peregrine jumps from a ledge to see the blue spheres' response: they save him and lower him to the ground, as a result of which he whispers, "You saved me! You wouldn't let me die. You knew it was wrong" (134).

- *Undeviating theology and its universal projection*: After all of the preceding, Peregrine instructs the other missionaries that they must proceed to build a church to attract the blue spheres in order to convert them, "to find their own special ways of sinning, the alien ways, and help them to discover God" (137).

- *Attention to the elaboration of alternative spiritual truth*: The blue spheres address the two missionaries, telling them that once they, too, had material bodies, but that a "good man" from among them "discovered a way to free man's soul and intellect . . . and so we took on the look of lightning and blue fire and have lived in the winds and skies and hills forever after . . . we shall never die, nor do harm. We have put away the sins of the body and live in God's grace. We covet no other property; we have no property. . . . We live in happiness. . . . We appreciate your building this place for us, but we have no need of it, for each of us is a temple unto himself and we need no place wherein to cleanse ourselves" (141–42). Note how the spheres "live" in the communal sharing manner of the early Christians in Acts 2 and 4, and have a variation of the spiritual understanding expressed by Paul in 1 Cor 3:16–17: people are temples of God. Note, too, that this indicates the blue spheres' expression of a higher level of both intelligence and spirituality, which are intertwined; this contradicts projections by contemporary atheist scientists that religion and spirituality will eventually be rendered obsolete in human evolution, having been superseded by advanced scientific knowledge and a higher overall human intellectual level. Here spirituality is not something *beyond which* humans will have evolved, but rather a higher level of consciousness *into which* humans will evolve.

- *Acknowledgment and acceptance of others' spiritual truth*: After their last, spiritually uplifting final meeting with the blue sphere Martian beings, Fathers Stone and Peregrine walk back to the human community. Father Stone, whose religious understandings were changed considerably by events during the expedition, observes, "The way I see it is there's a Truth on every planet. All parts of the Big Truth.

On a certain day they'll all fit together like the pieces of a jigsaw.... For this Truth here is as true as Earth's Truth, and they lie side by side. And we'll go on to other worlds, adding the sum of the parts of the Truth until one day the whole Total will stand before us like the light of a new day" (143). This complements Jesuit scientist Teilhard de Chardin's notion of cosmic evolution and convergence toward the Omega Point, which in his thought is the Cosmic Christ. (Teilhard's attempts to reconcile science and religion; his work as geologist on the scientific expedition in China in 1929 that discovered Peking Man, at that time the oldest known human remains; and his speculation about extraterrestrial intelligent life is discussed extensively in *Cosmic Commons*, and briefly in the present book in chapter 7.)

At the end of the *Martian Chronicles*, in 2057, a family tries to escape from Earth politics and destruction by traveling secretly to Mars in a private rocket. Hoping to start anew, they destroy their rocket and burn representative and symbolic papers brought from Earth: to maintain their resolve, to erase ties with the past, and to thwart pursuit from Earth, should any come. In the end, the father, who had promised his three children that they would see Martians, instructs them to look into the river by which they are camped: they see themselves. (If actual human colonizers who arrive on Mars in the future were to see themselves in this way and embrace their new socioecological context, humankind will have progressed significantly in consciousness and conduct from its mindset and its actions in the early twenty-first century.)

There is much more to be found in *Martian Chronicles* and others of Bradbury's works. However, the excerpts presented here suffice to reveal some intangible riches that might result from Contact in fact, as suggested or stimulated by considering presentations of Contact in fiction.

ETI on Screen

ETI in the movies or on television ordinarily represents one of two distinct and contrasting types: the *benevolent* alien or the *malevolent* alien. Some might transition from one to the other, such as the very different Klingon characters present in successive *Star Trek* television series (even being on board the *Enterprise* in later versions of the series), or Darth Vader in the *Star Wars* films that unfolded over the years. You probably are imagining, at the moment, both types of aliens whose memory stayed with you, perhaps for decades, after they struck fear or stimulated joy in their fictional roles: perhaps the lovable young "E.T." in the film of the same name, or the frightening physical appearance and aggression of Aliens in the series of films in which their name appears as the title.

In James Cameron's 2009 sci-fi 3-D movie *Avatar*, humans are extraterrestrials on the moon Pandora. In this setting, humankind thinks and acts just like we see ETI in sci-fi films that express our worst nightmares about extraterrestrials. We are the kind of ETI that we don't want ETI arriving on Earth to be: ruthless, technologically superior aliens who invade Earth to seize our planet and its natural goods, and commit genocide against humanity as a whole, as necessary, to displace us. In *Avatar*, humans do unto the Na'vi people what humans don't want others to do unto us on our home planet. We export and implement elsewhere in the cosmos the Discovery Doctrine that has wreaked havoc on Earth: physical and cultural genocide against indigenous peoples, and geocide and biocide against Earth and Earth's biotic community.

Cameron's *Avatar* and Bradbury's *Martian Chronicles* both portray humans as invading, dominant, and destructive extraterrestrials. We have "evolved" to become, we might suppose, an aggressive alien type dreaded not only by humans but by all intelligent species throughout the cosmos who have evolved to the extent of being concerned about invasion and the seizure of their treasured places.

Extradimensional Intelligence (EDI) in Extradimensional Aerial Phenomena (EDAP)

The prevailing understanding is that UFOs and intelligent alien beings originate from distant outer space. An alternative theory, proposed by astrophysicist and computer scientist Dr. Jacques Vallee, states that they are from a distinct but interwoven exoEarth dimension. Vallee's academic background includes a BS in mathematics from the Sorbonne, an MS in astrophysics from the University of Lille, and a PhD in computer science from Northwestern University. An astronomer at the Paris Observatory in 1961, he moved to the U.S. the following year to work in astronomy at the University of Texas, Austin. Subsequently, he went to Northwestern University and became a close associate and lifelong friend of J. Allen Hynek, who was at the time the scientific consultant for the USAF's Project Blue Book.

Extraterrestrials: Theirs and Ours

In the course of his extensive scientific investigation of UFOs and ET over more than fifty years, Vallee concluded that theories about extraterrestrial intelligent beings' voyages were inadequate to explain how they traversed vast distances and, in some narratives, how and why they disappeared instantly after accelerating to a velocity that seemed impossible to achieve under known laws of physics. He began to speculate and then theorize that what was considered extraterrestrial intelligence (ETI) might more likely be, in some events, extradimensional intelligence (EDI).

Vallee's extensive focus on statements of people claiming to be abductees, even though their recollections of their experiences ordinarily were recalled only while they were under hypnosis, might weaken his theory to some extent. While abductee stories have not all been discredited, nor has Vallee's overarching theory, his emphasis does provide an opening for those who reject his EDI theory. The latter have proposed or accepted the ETI hypothesis on the basis of physical evidence found during on-site examinations of places where reported phenomena occurred, and on sightings by witnesses at the scene that have been corroborated by airborne and ground radar. Credible witnesses who have reported visual sightings often are still unnerved by their experiences, even years afterward.

Evidence suggests that ETI and EDI theories need not be entirely rejected by their respective adversaries, nor exclusively advanced by their respective advocates, as if the choice were either/or. Both perspectives might be true, even to an overlapping extent. Some exoEarth travelers might, for example, begin their explorations as distant extraterrestrials with advanced technologies that allow for extremely fast journeys, perhaps through wormholes of some sort; or begin as ETI but then use an alternate technology that enables them to shift dimensions, as needed, to traverse vast distances in brief periods of time as EDI; finally, they might return to their origin dimension for the last stage of their voyage as they approach their destination. Their endpoint might have been predetermined, or only approximated if they sought to explore a new place in the cosmos and fixed their arrival point to a region that interested them. There are at least four possibilities, depending on their place and dimension of origin, for exoEarth Earth visitors' travel: ETI → Earth; EDI → Earth; ETI → EDI → ETI → Earth; and EDI → ETI → EDI → Earth.

Distinctive Designations

Concepts and terms that could designate and distinguish in an integrated way distinct types of exoEarth craft and intelligent beings, and allow for multiple possibilities of their origins and occupants, would include the following:

- *TI*: terrestrial intelligence
- *ETI*: extraterrestrial intelligence
- *EDI*: extradimensional intelligence
- *EEI*: exoEarth intelligence (both ETI and EDI)
- *TTI*: trans-terrestrial intelligence; human intelligent beings—TI—and extraterrestrial intelligent beings—ETI—originating within the same universe, who travel from one place to a specifically selected other place in that universe (with or without interdimensional segments of their voyage to save time), and explorers and colonizers voyaging among the celestial bodies of their own universe.
- *TDI*: trans-dimensional intelligence; intelligent beings who live in a different dimension of humans' home universe or from a different universe, who come into Earth's universe, or humans traveling to another species' world in a different dimension of Earth's home universe or in a different universe.
- *UAP*: unidentified aerial phenomena/on; aerial or seemingly aerial objects visible in or beyond Earth's atmosphere that transport ETI or EDI, directly or indirectly, or natural phenomena/on or human artifact(s) not initially recognized as such.
- *EDAP*: extradimensional aerial phenomena/on
- *UFO*: unidentified flying object; an ETI-operated (remotely or from within) spacecraft, or a natural phenomenon whose identity is not known or recognized.
- *UEDP*: unidentified extradimensional pheonomena/on; craft controlled by EDI and used to benefit EDI, and events that seem to have an exoEarth origin or source, all revealing extradimensional craft, intelligence, and technology.

- *UEO*: unidentified exoEarth objects; phenomena arriving from outside Earth whose origin and purpose are unknown: they might be UAPs or UFOs.

- *UOS*: unverified origin spacecraft; acronym to integrate Hynek's and Vallee's seemingly conflictive proposals on sources of exoEarth objects whose origin, purpose, and impacts are unknown, as based on their assessments of the same or similar evidence. It would refer generically both to ETI exoEarth craft arriving from a stellar region in the vastness of space, and to EDI exoEarth craft arriving from a different spatial location in a dimensionally distinct universe.

Enjoy playing with the acronyms—and looking for their manifestations on a clear, starry (are they *all* "stars"?) night.

Scientific Fact, Scientific Fiction, and Spiritual Formulations

What might ETI believe about the possible existence of a sacred Being in the cosmos? While this question will be explored in greater depth later (chapter 5), at this time a fascinating proposal about consideration of other Earth intelligent species' views offers insights worth pondering, for possible future application on Earth and in the heavens.

George J. Annas, a Harvard Law School graduate, attorney, bioethicist, and human rights advocate, some years ago took a time-out from his customary work as a university professor and human rights advocate to write a short story, "Dolphin Mission." It was published in *Analog Science Fiction/Science Fact* (July 1979). Scientific fact is that dolphins, perhaps for millennia prior to human evolution, were the most intelligent Earth species. In the sci-fi story Carlos, a Peruvian sailor shipwrecked on an isolated island and living by himself for decades, gradually learns to communicate verbally with dolphins, whom he finds to be highly intelligent. His sole possession on the island is a well-worn Bible. Carlos learns from conversations with dolphins Nina and Marco that members of their species believe that God, ruler of the planet, "is a great white whale that lives in a deep, distant sea." Carlos decides that God placed him on Earth to teach dolphins and other intelligent species about Christianity, "with the help of St. Francis." His mission seems confirmed when a young and bright dolphin, Patrick, accepts his teaching and is baptized. (It might be surmised here that Carlos intends that just as Saint Patrick was the missionary whose preaching helped convert Celtic peoples to Christianity, in a similar way Patrick will convert his dolphin people to Christianity.) Consequently, however, as he critically reflects on his intent in baptizing Patrick, he wonders if there are "some for whom the word of God was not meant" and who "can find happiness and peace and fulfillment outside the Gospels." He thinks, too, about how his ancestors captured and enslaved native peoples in the Americas, one impact of which was a genocidal smallpox epidemic. He wonders about natives' lives and culture before and after the Spanish arrived, and how that might relate to his missionary efforts among dolphins.

Shortly thereafter a dolphin carries to shore another sailor, Jack. Carlos decides that Jack has been saved from the sea, like Jonah in the biblical story, to help convert dolphins. Jack wants no part of this. He doesn't believe that dolphins are intelligent and, even if they are, he has a right to kill them for food (even though they have been providing fish for Carlos, and now him), since all creatures are subject to humans as taught by Genesis 1. In a subsequent conversation, Jack tells Carlos that his interpretation of Jesus as Savior is unreal, that Carlos wouldn't recognize Jesus if he walked across the ocean to Dolphin Island, and that Jesus might even have been "nothing more than some misguided traveler from another solar system, bringing a message we weren't ready to hear." Eventually, Jack lures Nina to shore by feigning friendship—and kills her. Enraged, Carlos kills Jack. Paradise is lost.

"Dolphin Mission" suggests intriguing possibilities, directly and indirectly, two of which are that other intelligent Earth species, despite religions' teachings, might have their own understanding of and relationship with divine Being; and that, extrapolating this insight to or complementing it with independent speculation, it could be acknowledged that exoEarth intelligent species, too, have evolved to have (or have culturally developed) spiritual traditions based in and related to their culture(s). The latter would enable them to relate, as individuals and species, to a sacred Presence mutually engaged by all cosmic beings to the extent of their capability to do so.

In an analogous sense, Carlos was ETI to the dolphins' TI. When he was shipwrecked in the water and after he was carried to the island he was "extra" to their territory (he came from an external territory where his home was located), to the vast space and places they had inhabited for millennia. Gradually, he acclimates and acculturates and becomes an individual who is part of an additional area TI species. His reflective questions, like some of the ones posed by the priests in *The Martian Chronicles*, reveal his gradual acceptance of what was at first to him an "alien" culture. (Historically, it might be noted, his Peruvian origins likely meant that he was a mix of Spanish and Peruvian *indio* ancestry and cultures. His developing human-dolphin consciousness enabled him to be a parallel bridge between species cultures.) His religion, as expressed in his Bible and interpreted by his church, becomes a barrier to full acceptance of and by the dolphins. (Similarly, the Discovery Doctrine Spaniards had their "true religion"—Christianity—and church—Catholicism—and were not open to the spiritual insights of *indios*.) Jack, by asserting human superiority to all creatures, no matter their intelligence, affirms his ongoing ETI status; when he murders Nina, he confirms that for the dolphins and Carlos. Unfortunately, Carlos reverts to his ETI culture consciousness and thereby status when he kills Jack. The TI dolphins then refrain from providing him with food, fearing to nurture him. Patrick secretly comes, tells Carlos he is a dolphin scientist who has been studying humans (his apparent acceptance of Christianity was to further his research), and offers to transport Carlos, undetected, to a distant island where he will have sufficient supplies to provide for his needs. Carlos agrees, under the condition that Patrick stop referring to him and Jack as "Cain and Abel."

Like Carlos and Jack, ETI might be benign or bellicose, or both, within a single species or as distinct species. Perhaps both types of intelligent life exist on different planets and different spaceships throughout the cosmos. Perhaps, alternatively, an intelligent species has elements of both, and uses its selective adaptation ability to respond as needed to new contexts. Perhaps, too, they have come to uneasy accommodation or adaptation when they come into Contact. Perhaps, instead, intelligent species' integrated biological and social evolution on their home planet, if not impeded and stymied by an immature, static, and self-centered cultural consciousness and consequent immature conduct, leads eventually and even inexorably, during their explorations, toward global harmony and wellbeing and thence to cosmic curiosity and congeniality, rather than combativeness and conquest. They might be disposed to be collaborative, as a consequence, when they Contact other intelligent beings.

In light of such uncertainty, and projecting onto others as one possibility our own immature consciousness and conduct, however erroneously we might be doing so, we might well be prepared for either possibility, both possibilities, or multiple possibilities. For their part, ETI of varied species might already be observing *homo sapiens* wondering if the *sapiens* part will be freed from individual and cultural constraints and achieving the wisdom needed to have harmony and mutual wellbeing on Earth and be welcomed among the older civilizations that have traversed the stars long before Earth came into being. That thought, too, can inspire and guide us at the dawn of the Star Age. We would, in that case, strive not only not to destroy ourselves back to the Stone Age in one of the ways about which Stephen Hawking and others are concerned, but to become one integrated, interdependent, and interrelated human family on Earth. Then we would be among the stars, in our exploration and settlement, the kind of extraterrestrials that we hope ETI will be when they step onto *terra firma*.

Malmstrom Montana Minuteman Missile Meltdown

Now that was a mouthful! But it caught your attention, especially if you know what an ICBM is and that the Minuteman missile is an ICBM. For the younger reader, or those absent when this topic came up in high school and college U.S. history classes, ICBM is the acronym for intercontinental ballistic missile, the U.S. version of the long-range nuclear missiles that the U.S. and U.S.S.R. kept individually cached in underground silos, ready to be launched within minutes from when the command is given to destroy the enemy nation, or selected parts thereof.

Montana has a wing (150 missiles) of ICBMs, divided into fifteen flights (10 missiles each), all under the overall command of Malmstrom Air Force Base near Great Falls. Each flight has its own missile control center, a capsule sixty feet underground, in which two officers ranked lieutenant or higher who have Top Secret/Sensitive security clearance are on twenty-four-hour shifts. During their capsule period, they take turns sleeping, eating, etc., while their fellow officer stays at the console that controls ten missiles in separate missile silos spread over an extensive area. Each missile has its own switch for launch, and a light that stays lit to indicate that the missile is online and ready. There are at least three backup systems on-site, and three others at Malmstrom, which is located more than a hundred miles from several silos. The backup systems are supposed to enable missiles, individually or collectively, to be restored to online status should a light go out to indicate they are not operational. On the surface above, the command center and its own missile silo are protected by an armed squad of Malmstrom air security police personnel.

On March 16, 1967, when Lieutenant Robert Salas was on duty at the Oscar Flight missile control console, he received a phone call from the leader of the on-surface security police contingent. Salas was asked what the squad should do regarding an object that had flown strange patterns in the sky, then descended to hover above the silo site. Salas didn't take him seriously, told him to secure the perimeter, and returned to watching the console. Within a few minutes the security guard called again, his voice frantic this time, shouting that one of the guards was bleeding and the object was still overhead. Concerned now, Salas went to the surface, saw a large dark object directly overhead blocking the stars, and the bleeding guard, and descended to the control center. Strangely enough, he reported later, he felt no fear because he sensed a reassuring telepathic communication telling him that he shouldn't worry. Once below again, he woke Lieutenant Fred Meiwald. As they discussed surface events the alarms in their capsule that were linked individually to each of their ten missiles sounded loudly, singly and in clusters, until each and finally all of the ten missiles went offline while the UFO was above. They tried backup systems, but the individual lights remained unlit; they couldn't restore the missiles. System checks indicated that power had not been lost. Rather, each had a Guidance and Control (G&C) System fault. Since this was a time when the Vietnam War was raging, and the Cold War was heating up, loss of missile controls was a serious national security issue. When he returned to Malmstrom, Salas learned that a similar event had occurred days earlier at Echo Flight, close to the Canadian border.

There were other reports about UFO impacts on ICBM systems in the U.S. in subsequent years. There might have been several incidents in the former U.S.S.R., only one of which has come to light. Retired Army Lt. Colonel Vladimir Plantonev told ABC news in an October 1994 interview that in 1982, in the Ukraine, a UFO flew for several hours in the skies above an ICBM installation. During this time the missile launch system was suddenly activated for fifteen seconds, then restored to its ordinary operational status, and then shut down.

The Malmstrom and Ukraine events are elaborated in greater detail in *Faded Giant* (1995), by Robert Salas and James Klotz, and *Montana UFOs and Extraterrestrials* (2012), by Joan Bird. Such incidents indicate that ETI visitors understand well the political dynamics and mutual hostility of nuclear nations, their respective ICBM nuclear weapons technologies, and how to control them—and ETI shut them down. No large-scale invasion followed, nor even a focused attack. This might be an indication of peaceful intent declaring, in effect (and effectively), "You're on the wrong track, don't destroy yourselves." It might indicate also that ETI—at least some ETI visitors to date—has evolved to be benign, but probably is defensive, when and if needed, if it is attacked by hostile intelligent life, or confronted with hostile activity from frightened intelligent life.

I've Seen a UFO: What Should I Do?

Well, probably the first thing would *not* be to call the government. Whoever responded would tell you that you did not see what you knew you had seen. Remember, at the present moment in history only a select few people know for sure what was recovered and secretly stored in the aftermath of the Roswell crashes of two or three spacecraft, as seen and reported by credible witnesses. Even after unauthorized disclosures of what remains of alien beings and remnants of alien craft have been hidden by the U.S. government, and what files have been kept, the unabashed government denial has continued. In a departure from this policy, retired Lieutenant Colonel Philip J. Corso published *The Day After Roswell* (1997). The book describes Corso's direct contacts with alien remains: he saw an alien body recovered from Roswell when it was being shipped through Kansas to Baltimore and, years later, while stationed in the Pentagon, saw the autopsy report on that alien. Seventeen years after Corso's book was published, the U.S. government continues to deny that there were Roswell area UFO crashes, that it has debris from them, and that there were alien bodies recovered. Meanwhile, Colonel Dr. Jesse Marcel, Jr. has been trying to prod the government to restore his father's good name and professional reputation, both of which were impugned when he was forced to remain silent about alien debris he found at Roswell.

Paul Hellyer, former Canadian National Defence Minister and member of the Canadian Parliament for more than twenty years, said in email correspondence with me in August 2013 that, regarding witnesses of UFO and ETI events, "I would encourage anyone who has knowledge that should be in the public domain as a contribution to the general long-term welfare of society, to go public with what they know. This despite the fact that there will be a price to pay. The risk applies to pioneers in any field whether it is someone who talks about their breast cancer, their problem with depression, an athlete admitting that he or she is gay, or someone talking about their knowledge of the extraterrestrial presence or technology. Certainly Dr. John Mack [of Harvard University] paid a high price at his university but is now recognized by people in my circle as one of the outstanding men of his time."

So, when you see a UFO that appears to be an alien space vehicle, shoot a video, or at least take a picture with your camera phone or a camera if you have one handy, and contact a local newspaper. If the UFO is hanging around, suggest to a local TV station that they come to where you've seen the spacecraft (today, more than in even the recent past, news media are willing to send a crew to record or at least report on such an event). Do *not* give your only copy of the footage you have, if any, to anyone—even if they claim to be from a U.S. government agency, a branch of the military that needs your personal media property because of "national security" concerns, or a member of the news media. Keep the original for yourself, and make a copy or copies for others. Keep in mind the story of Nick Mariana, a college graduate, WWII vet, and then general manager of the Great Falls Selectrics minor league baseball team. In 1950 he videotaped two UFOs flying high above the tall Anaconda Mining Company smokestack near Great Falls. He submitted his only copy of the videotape to U.S. government officials to be examined. It was not returned to Mariana for a while until after he made numerous requests for it. However, the returned tape was not intact. The clearest and most dramatic footage of the flying UFOs had been removed, and the video had been spliced. Fortunately, Mariana had shown it to staff from the *Great Falls Leader* newspaper and to his local Lions Club at their weekly meeting. Before the Lions meeting began, members had joked about what he was about to present, and even hurled paper plates across the room as "UFOs." As the footage ran, however, they fell into awed silence at the sight of the two objects hovering over the Anaconda Company smokestack. In 2008 the Great Falls White Sox baseball team changed its name to honor Mariana: now they are the Great Falls Voyagers, whose logo is a cute alien named "Orbit"; he has the requisite lightbulb-shaped head and large eyes. This is especially appropriate in this city for an additional reason: Great Falls' nickname is "The Electric City." (Joan Bird's *Montana UFOs* has additional details.)

Highly respected sites whose staff you might contact include CUFOS (J. Allen Hynek's Center for UFO Studies); CUFON (Computer UFO Network); MUFON (Mutual UFO Network); and theblackvault (which mines government documents for UFO/ETI references, in part by use of Freedom of Information Act requests).

To Believe or Not to Believe: That Is the Question

Well, not for everyone. Some are irrevocably committed to one "side" or the other, and will continue to be unless some compelling strong evidence alters their belief. Others have seen what they understand to be UFOs, and so they need not believe: they already "know": they are beyond a belief in something unseen. If you're still an "agnostic" on this issue, finding no compelling evidence or reason to believe (such as apparently credible witnesses, whom you might think are mistaken in what they testify to having seen) or not to believe (because of widespread dismissal or derision of ETI and even UFO reports—even when accompanied by photos or video footage, which are always claimed by "disbelievers" to be fabricated, something made increasingly more easy because of current technologies, available even on home desktop computers, to manipulate or alter photos, movies, etc.). Perhaps, too, there's a psychological reason why you *choose* not to believe, despite what seems to be evidence of extraterrestrials: fear of the unknown; fear that you'll realize that humans are not the most evolved and technologically advanced species in the universe; fear of weapons that ETI might have and use; fear of impacts on religious faith if some comfortable, long-held beliefs seem to be compromised or negated by ETI existence on distant planets or other places in the cosmos, perhaps since billions of years before Earth and its solar system came into existence, and therefore well before human evolution even began; fear of reactions, maybe including ridicule, from family, friends, professional colleagues, employer, or others; or, any combination of the foregoing. Or, you might have a professional fear about others' knowing about your UFO/ETI interest even if it's only your innate curiosity that prompts your questions and speculation: your colleagues or employer might not only ridicule you and your ideas, but even cause you such harm as loss of job, demotion or lack of promotion, diminished salary, etc. The U.S. government and academic institutions beholden to it (research funding would be jeopardized if your pondering is not ridiculed and suppressed) might be a factor, until a sufficient body of academics courageously exclaim "Enough!" or "I'm not going to take it anymore!" and advocate for scientific research by independent bodies to analyze UFO data objectively. They might then theorize, if not establish definitively, the causes of UFO phenomena, and the nature and technological capabilities of those who are responsible for making and operating them.

To Research or Not to Research: The Real Question

It seems reasonable to objective, intelligent, curious people that the tools and teachers of science should assist in analyzing UFO accounts. Science is not as "objective" as some think it is (note the bias against UFO studies noted above, and the experience of Allen Hynek). However, properly monitored and independent science research, undertaken in field and laboratory by scientists who have established reputations and would not want to have them called into question because they steadfastly hold to a fixed position when evidence to the contrary becomes studied and accepted by other distinguished scientists, would play an important role in UFO and ETI data analysis. They could suggest core hypotheses in order to continue to clarify such matters.

Several possibilities come to mind concerning types of people and institutions that might courageously engage in such endeavors: individual independent scientists who are financially secure and can fund their own research; nonprofit scientific research corporations whose funders—individuals, foundations, and corporations—are sympathetic and supportive of efforts to "find the truth" about UFOs; academic institutions and specific science research departments that might have instruments and staff available—indeed, willing and able—to do this work; "think tanks" organized for this specific purpose and able to secure funding from foundations governed by intelligent, curious, and wealthy individuals who have a passion for research and knowledge, and are (perhaps secretly, to this point) eager to explore the issue in depth, and have the funds to do so. Where *are* the courageous foundation staffs and boards who could and would fund scientific research into potential extraterrestrial intelligence-related matters if they could find intellectually courageous academic institutions and independent research organizations and institutes that would undertake it? In all these possibilities, it might be deemed politically and socially necessary to keep matters secret initially, and gradually build up a dedicated core and then a network of professionals interested in a serious scientific quest that is dedicated to finding facts and disseminating them to the general public.

People are hungry for objective, accurate information about UFOs and ETI. Who will step forward to satisfy this hunger? Which readers of

Encountering ETI experience this eagerness for truth, and for accurate assessment of UFO events and data? People from Roswell, the Hudson River Valley and its environs, and Rendlesham Forest who understand that they have seen evidence of ETI, are yearning for others to share in this understanding and be supportive of their efforts to stimulate awareness of data and facts among members of the general public. Surely people who are objectively skeptical would want this, but not people who tend to believe their government "no matter what" if it's led by their particular politicians.

To Know or Not To Know: The Most Important Question

When all is said and done, we humans prefer, to the greatest extent possible, to have accurate knowledge about whatever is important to us, or whatever is most piquing our interest and curiosity. For many of us in the U.S., knowledge about UFOs and ETI, particularly since it has been augmented already by official reports from government officials, military leaders, and scientists in other nations, ought to be supported by the U.S. government, academic institutions, and independent scientists; all should be transparent about the UFO and ETI knowledge they already have gathered and secreted into files, and knowledge that continues to come in. When will the "free press" that supposedly exists in the U.S. undertake serious investigations, beyond reporting an occasional story, into UFO and ETI matters? Surely some reporters and editors would welcome such work. Perhaps a Pulitzer Prize might result for the writer of an excellent, in-depth, well-researched original story in this area of inquiry. Well?

A question or questions are generally required to attain knowledge. The question might be asked of a teacher or posed to oneself. If the question is suppressed by the one who has it, because of insecurity about implications of the response they might receive, they will not benefit by clarifying their current knowledge or acquiring additional knowledge. Effectively, they have self-censored—to their detriment, their peers' detriment, and their mentor's detriment. Over the years, in the first session or two of my college undergraduate courses or my university graduate courses, I have instructed my students not to be afraid to ask questions they have as a result of required readings for the course, or of material presented in my lectures. I realize that sometimes students who are studying a new topic, or an old topic in new ways, will be afraid of appearing "dumb" in the eyes of classmates whom they assume (because of their corresponding silence) to have understood the text and the lecture. I ask them to consider the following: it is possible that several or all students in the class have the same question, but all are afraid to ask it because of how their peers might view them; it is possible that no one else has the question—neither their peers nor the professor—because they have not had a similar insight or perspective (perhaps because of a different educational or experiential background), and when it is not asked all who are in the room and their extensive contacts lose the knowledge that it would have generated or the quest for a response that it would have stimulated; it is possible that the question might catalyze a whole new area of inquiry. All of this is lost, at least for this potential creative moment, because people do not ask their particular question(s).

My students' fears are analogous to professional people's fears that no one else has their questions about UFOs, and they do not want to appear to qualify for social or intellectual dismissal, or ridicule by others, should they express either their curiosity or their experience—whether direct, or indirect through credible witnesses—with UFO events. On several occasions, I have presented a lecture or made a casual reference to my UFO/ETI interests in a university setting. Very rarely has any of my listeners publicly expressed a similar interest or experience when we are in the group as a whole. However, afterwards colleagues from diverse fields have approached me and quietly described their own, similar experiences and their secret hope to be able to communicate them, at least to friends or family.

To give an example, several years ago I sat in a group of four colleagues in the casual setting of a faculty social, a place and time for people from across the university to meet others for mutual personal and professional benefit. We discussed summer plans. I said that I was going to offer a course on "Encountering ETI." There was somewhat shocked silence: such matters are not discussed in a research university; only the ignorant raise the topic. However, after I described my experience in the Hudson River Valley, one of the two very senior research scientists who were present and had dismissed the UFO possibility, described experiences he'd had in the 1950s while stationed on the Maine coast as an airplane spotter for the U.S. Air Force. He was to carefully watch the skies over the Atlantic Ocean looking for invading U.S.S.R. airplanes. On one occasion, he saw flying objects approaching in formation at an extremely high speed. He studied them carefully and knew they were neither "ours" nor "theirs." When he called his base he was told the first time that radar was not detecting anything; the second time, that he must be mistaken in thinking that they were neither

U.S. nor Soviet military planes, an accusation that he rigorously and angrily rejected; and the third time: "You did not see anything. Repeat: you did not see anything." He knew very well that he had just been given an order that he was to obey. This story greatly surprised his scientific colleague and close friend who had never heard him speak previously about this event. Then the fourth person described an experience had by her "conservative Kansas aunt." She had told only selected family members, in confidence, about the UFO she saw outside her kitchen window as she washed dishes: it landed in the farm field, close enough to be clearly identifiable, but unlike anything she had seen before and clearly "not from this Earth." She, too, feared social stigma or at least ridicule—not only from other farmers and rural community members but also from her own family—if she were to talk openly about what she'd seen. So, three out of four university faculty members who met casually and coincidentally at an informal gathering had had some experience, direct or indirect, with UFOs. Two had not dared to discuss such events openly previously during their years at the university, or even in their own off-campus social circles or family gatherings. My narrative opened the door for two colleagues whom I had not met previously to finally express to others, with some evident relief, what they'd experienced but had not dared to acknowledge publicly before—even to family and colleagues who were friends.

In 2013, when I was casually conversing with two distinguished Boston University faculty friends about our respective summer plans, I mentioned that I was working on this manuscript, and that my other book on TI-ETI Contact, *Cosmic Commons*, was in press. They looked first at each other, then at me, and one said, "You'd better not tell anyone on campus about this." In this case, my friends were not worried about what would happen to themselves personally or professionally if they mentioned UFOs or ETI, but were concerned about repercussions for me and my career. I am grateful for their solicitude. I'll be more careful on campus, perhaps—except for select friends—but not when I'm out of state. In any case, anyone at BU not aware of my interests will know soon enough when the books are published.

There are hundreds of other professionals, it seems reasonable to assume, some of whom are reading *Encountering ETI*, who have had similar experiences but dare not disclose them to others. Who of you will dare to quietly talk with family or friends about your experiences or curiosity, perhaps at a cocktail party or over dinner where all present have, over the years, become comfortable in each other's presence and mutually respectful? You might be pleasantly surprised to learn from others there, perhaps even after several people present have expressed disbelief or made jokes about "little green men," etc., that one or more of them had similar experiences or are at least curious about UFO reports. Or, someone might call you afterward, or stop you at work, to mention—with pleas for anonymity—that they, too, have had a close encounter, perhaps through an astronomy department telescope or in some more direct manner. (Perhaps *Encountering ETI* might serve as a book group's or other group's common reading to promote discussion of UFO/ETI topics.)

I still note that some people smirk or roll their eyes when the topic arises. Some seem to have the attitude that they will not accept UFO narratives presented by others: they will not "believe" unless they "see the evidence" themselves (though they do not have a similar attitude toward people in their own field or toward the authors of the texts they use in courses or people cited in them), or they are insecure in their own field and do not want to consider data or stories that would appear to threaten their long-held beliefs. Yet others are open to think in new ways regarding UFOs and ETI, despite the possible unsettling of their prior beliefs, ideology, and even supposedly hard data.

Ask questions. And share your own experiences and knowledge.

3

ETI: ROSWELL, RIVERINE, AND RENDLESHAM ENCOUNTERS

We Are Not Alone . . . ?

THERE HAVE BEEN NUMEROUS narratives about human Contact with extraterrestrials throughout human history. During the twentieth century, stimulating stories about direct (sightings of exoplanetary beings) or indirect (sightings of UFOs) Contact between terrestrial intelligence (TI) and extraterrestrial intelligence (ETI) throughout the U.S. have meandered through the media. Three such sites of TI-ETI events, as described by those who experienced them, will be considered in depth: Roswell, New Mexico; the Hudson River Valley in New York and neighboring areas in Connecticut; and Rendlesham Forest, England. The reasons for the limitation of consideration to only three places are space and time constraints; the extensive body of credible evidence and analysis available regarding incidents in the locales mentioned; and the decision to accept and not to question or debate the veracity of credible witness accounts (after all, they are "credible witnesses") or the data expressed in them, but to use the narratives to focus on potential implications of terrestrial-extraterrestrial intelligent life Contact. Thus, we will ponder together these three events as part of our "thought experiment" in *Encountering ETI*: the incidents will be discussed "as if" they describe actual historical events and human experiences. This will be done so that we'll be able to consider more in-depth what reactions there might be in the future if Contact occurs/recurs. Our continuing purpose is to ponder impacts and implications of actual or potential TI-ETI events on humankind—on Earth and in the cosmos beyond.

Some people scoff at the idea that ETI has visited Earth at any site; others believe or know that this has occurred, from direct experience or through the credible witness reports and statements of others whom they deem trustworthy. The Roswell, Hudson River, and Rendlesham events, seen through the eyes of actual witnesses and thereby viewed as historical, will help us review critically what they might mean for humankind in particular. Our ideas and reflections then will help us—whether or not such events have occurred or will occur—to establish a dialogic relationship between Earth and exoEarth places in the cosmos, between ourselves as TI on Earth and the ETI we might Contact here on Earth; and, the TI we might Contact on other worlds when, as ETI, we arrive there and encounter the existing TI inhabitants (thinking of them as "terrestrial" in such moments, since we would be arriving from off-planet onto their "terra firma"). We might also come to be in dialogic relationship between present and potential realities of our contextual cosmic presence in distinct settings—and our developing sense of the meaning of our presence—as we begin our space ventures and adventures, and even beforehand as we ponder relational possibilities.

The suggested thought experiment will stimulate us to consider proactively what we might come to experience at Contact moments. Somewhat suddenly, even with such a preparation, we might have a sense of displacement from our self-designated position of superiority in the cosmos, a status to which some of us might strive to cling—sometimes, perhaps, to the point of being overwhelmingly in denial when confronted by narratives or even evidence to the contrary. As a consequence of irrefutable proof of ETI, we would have to thoughtfully reconsider what our place might be as intelligent members of the totality of integral being in the cosmos.

When we focus on ourselves in our primarily pre-Contact setting, think about ourselves as soon-to-be-voyaging in space ETI, and look at what we've been doing on Earth to our home and habitat—the common ground we share with all biota—and how we've even oppressed and continue to oppress members of our own species, we might be stimulated to do better elsewhere, whether or not we encounter "TI" native to other "terras." We might also have one of those "Aha!" or "light bulb" moments and wonder, "Shouldn't we have, and why don't we have, a greater sense of responsibility to care for our home planet and all members of our species—let alone of other species—as we speculate what we would be like if we were to venture into places where other intelligent life exists and thrives, or even where it does not exist?" Remembering Stephen Hawking's comments, we should resolve that we don't want to do elsewhere—and will not do elsewhere—what we've been doing on our Earth home to our environment, to our ecology, and to vulnerable populations at different stages of technological development. So, we act "as if" these UFO and even ETI narratives are historically accurate or predictive of future events because even if, beyond mathematical probability, we are the only intelligent life in the cosmos, we should take better care of Earth. If we do so, our solicitous care of other places and "peoples" in whatever cosmos contexts our journeys will take us to might be assured.

TI-ETI Contact: *Homo sapiens sapiens* Displaced?

Cosmic dis-placements—events or experiences on a cosmic scale that affect our ways of thinking such that they dislodge us from our perceived sense of who we are and what our status is in our world and universe—have occurred before. Once upon a time in human history, specifically the history which included the origin area for the three major religions—Judaism, Christianity, and Islam—and eventually what became known as Europe, people regarded themselves as the alpha species in the universe. We were superior to and enthroned above all others, the only biota with intelligence, specifically created by God to be rulers of a world which was created for us. This world was believed to include an outer shell, a "firmament"—a hard dome on which sun, moon, and stars orbited around Earth. So, our planet was the center of the cosmos, a world that was decidedly, by divine intention, anthropocentric—human-centered. By extension, this made humans the center of the universe, around whom everything revolved. In Christianity, this was strengthened by the belief that God became incarnate—enfleshed—on Earth to save people from themselves because they had stopped following God's commandments. It was believed that God became part of God's creation only in one place, Earth, and only as a member of one form of life, humankind, and only in one person, Jesus. Eventually, the focus on following the *teachings and example* of Jesus—early Christians called it "the Way"—was transformed into following the *cult* of Jesus (religious rituals, and religious beliefs institutionalized into dogmas, all of which was conceived and overseen by clergy or those whom they designated, rather than developed in and affirmed by community): *orthopraxis* (right conduct) became subordinate to *orthodoxy* (right belief).

Let's review the preceding: humans believed themselves to be the center of Earth, the center of the universe, divinely placed over Earth and all other creatures (all of which were created for humankind), specifically created by God from Earth's earth and in some way an "image of God," the only intelligent species, and the inhabitants of the only habitable world, which was at the center of the universe. Understood?

Keeping this in mind, and placing yourselves mentally in the age of Galileo (late-sixteenth- to mid-seventeenth-century Europe), imagine what your reaction would be if someone dared to suggest that Earth was *not* the center of the universe. That would mean, of course, that you were not at the center of the universe either. In fact, your world was not even the center of the system that included a star and planets: it was *heliocentric*, sun-centered, rather than *terracentric*, Earth-centered. You might be shaken up a bit. Hastily, back then, the Catholic Church required Galileo to retract his scientific findings (which had been expressed before him by Copernicus), and for most people their understanding of their world returned to "normal." Eventually, however, the overwhelming evidence for what became known as the Copernican Revolution was incontrovertible, and people came to realize and accept that Earth was indeed just one of several planets that orbited the sun. Of course, if you lived elsewhere in the world where such discoveries had not been made, and word had not arrived via traders from Europe, you would continue to believe what the ancient Israelites believed from observation: it's pretty obvious from what you see that the sun "rises" in the east, orbits Earth, and "sets" in the west. Even in highly industrialized nations today, most likely your local television station's weather forecast each evening provides a bit of data that you know is presented with inaccurate phrasing: the times of "sunrise" and "sunset" for the following day. Here's an experiment for you, involving a shift in focus that becomes a shift in both physical (eyesight) and psychological ways of seeing. If you're a morning person, without endangering your eyes, look at the sun and Earth's horizon differently: instead of seeing the sun rise, keep the sun fixed and watch Earth's horizon "fall" ever lower, and ever more distant from the sun's disk; you're seeing Earth's rotation on its axis as it orbits the sun. Similarly, in the evening: keep the sun fixed and watch Earth's horizon "rise" ever higher until it hides your view of the sun: again, you're seeing evidence of Earth's rotation. (You might come to realize too, as a result of this experiment, that there are other areas of your thinking that should have a shift in perspective in order for you to relate to reality more accurately and holistically.)

Stay with me, now: there's a reason for this much-abbreviated history and science lesson. You've now become accommodated to Galileo and no longer want to burn him at the stake for heresy. (In fact, the Roman Catholic Church retried him in the twentieth century and found him innocent of heresy, thus belatedly somehow displacing its earlier trial and correcting its centuries-old error.) Move ahead about five centuries in your mental journey. It's now 1860 and by this time you've heard (educated person that you are) that in the previous year a scientist in England published an outrageous book, *On the Origin of Species*.

In his book, based on his initial studies as he traveled around the world as a natural scientist on the HMS *Beagle* and on his subsequent research, Charles Darwin proposed his theory of evolution by natural selection. He stated that over millions of years, humans had evolved from a primitive life form into primates, and then into a distinct human species that emerged from ancestral apes. "What!" you might have exclaimed. "I'm not a rising ape: I'm a specific creation of God. The first book of the Bible, Genesis, says so in two stories in its first two chapters." Gradually, scientists found fossil records of the evolutionary transitions from one species to another, including dramatic ones in which fish fins over time become appendages that enable some species to walk, to climb out of the water and onto land, and in which some dinosaurs evolve wings and learn to fly. As a centuries older representative human being now, you realize that just as when you accepted that the sun did not revolve around Earth you had to recognize that the Bible is neither a history book nor a science text, now you must recognize that you are not a special divine creation directly placed by God on Earth, but a species that emerged from a preexisting species (you might partially accommodate the new and the old by believing that while your body evolved from apes, your soul and that of every human was individually created by God). As a representative human from a particular culture, who is gaining increased knowledge about Earth and cosmos (all of which constitute creation for the theist) through centuries of time, you've now experienced your second displacement: humans were not physically special divine creations: your specific materiality emerged through billions of years of biotic evolution.

A new possibility could catalyze in a dramatic way your third displacement as it hovers on the horizon (and perhaps even closer than that). As a representative human in the twenty-first century, you might suddenly be confronted by definitive proof that other intelligent species exist in the vast cosmos. This would indicate that "you are not alone" in being intelligent. In fact, if intelligent species evolution is a natural (inevitable?) process on planets once life begins, then since Earth is about 4.5 billion years old (and life on Earth is about 3.5 billion years old) while the cosmos is 13.8 billion years old, and since the Kepler and other telescopes have found planets similar to Earth that circle other stars, it is very possible that intelligent species Contacted today would be billions of years older than we are, and much more advanced technologically. This might result in a different experience and second aspect of humans' third displacement. Humankind would be dis-placed not only from its perceived place in the cosmos but also its actual place on its home planet: a direct *physical* (terrestrial-geographic, political, economic, habitational) displacement. This would not happen if only a *psychic* (cosmic-psychological and spiritual) aspect of the third displacement were to occur. It could be a frightening prospect—as some science fiction films portray very graphically. On the other hand, if stories about Contact to date are accurate, it seems that most ETI are more curious than contemptuous of us, perhaps hopeful of cultivating a congenial relationship with us, and uninterested in conquering us.

In any case, most of us might have a certain conscious or subconscious fear about the possibility of our third displacement from the status and role our forebears believed we had, and of the status and role we think we have today. An example of humans' self-perception and self-interests regarding exoplanetary regions is found in the United Nations 1967 Outer Space Treaty. It declares that the exploration and use of outer space "shall be the province of all *mankind* [*sic*]" (emphasis added). The statement was included because of international concern that the U.S. and U.S.S.R., then the only nations with atomic weapons capability and space exploration technology in its earliest stages, would seek to claim for themselves any planets or other "celestial bodies," in U.N. parlance, where they arrived first. This would continue the practice in space of the Discovery Doctrine used on Earth by European nations from the fifteenth to the seventeenth century to claim sovereignty over lands in the Americas and Africa, even though these territories were already occupied. The twentieth-century U.N. representatives had no consensus concept of nor concern about ETI when the treaty was written. As ETI existence becomes more possible, at least mathematically and because of astronomical observations, some provisions of and declarations in the Outer Space Treaty become obsolete: not only because they do not recognize the intelligence and potential needs of ETI for territory and natural goods, but because ETI is very likely more technologically advanced—including militarily, in terms of weapons at its command—and would oppose human claims to the entire cosmos. The "superior civilization" component of the Discovery Doctrine would obviously not apply to TI, but indicate the interspecies social status and standing of ETI. If ETI were to have such a Discovery Doctrine, and thereby an unjust theory, consciousness, and conduct similar or identical to that of Europeans and Euroamericans, they could use the Discovery rationale when they discover Earth and humans.

Some people today cannot accept the second displacement: they continue to regard the Bible as a history book and science text, and to take literally the biblical stories about cosmic and biotic origins. They believe that Earth is little more than six thousand years old, humans and dinosaurs lived simultaneously, and all biota are special, direct divine creations. Some modify this, since the evidence of some species' evolution is compelling, and state that while other biota evolve, humans are God's special creation. Many think, too, that if there is intelligent life elsewhere, and since Jesus came to "save" the "whole world," ETI will have to be baptized—for their own eternal benefit. Therefore, they would not accept the first aspect—the *psychic*—of the third displacement. They believe humans have a higher status than other intelligent beings because God became part of God's creation in only one place in the cosmos, at only one time, and as a member of only one intelligent species, humankind. As a result of this divine choice, humans remain superior to others: God became one of us, not one of "them."

In light of what we've discerned over time, and as you've seen over the centuries as an ageing and still reflectively thinking human representative who has now arrived in the twenty-first century, you can understand well why some people strongly resist "believing" in UFOs (narrowly defined as a craft controlled or crewed by ETI) and in ETI. Their fundamental beliefs are being challenged as their psychological fears about biological inferiority, their perceived loss of what they had assumed to be their planet's status, and their fear about their possible displacement from their physical place, all are exacerbated.

Atheists, of course, have their own faith systems, which include their belief that there is no divine being (whether called "God," "Yahweh," "Allah," "Wakantonka," or by any other name) and that materiality defines and limits reality. They would want hard facts before they would "believe" in UFOs or ETI. They too might well be consciously or unconsciously fearful about being less intelligent than another species, or about planetary and cosmic displacement. At least they wouldn't feel compelled to "baptize" ETI—they don't believe that ETI is religiously inferior and in danger of damnation (or, of course, that ETI is religiously superior—they think that ETI would have evolved beyond a "need" to have a religion).

UFOs: Exciting News and Photos!

A compelling case for UFO existence and encounters is made by former National Public Radio investigative journalist Leslie Kean in *UFOs: Generals, Pilots, and Government Officials Go on the Record* (2010). Kean uses the Freedom of Information Act to try to force the U.S. government to disclose the contents of its secret UFO files—which it claims are nonexistent. Kean also presents information from global sources—including from ordinarily reticent people as listed in the subtitle, and from some ordinary citizens who were previously fearful of retribution from U.S. government officials and agents striving to suppress UFO/ETI information. It includes, too, high-quality, extraordinary UFO photographs provided by military officers from several countries. I highly recommend Kean's book: it provides an insightful overview of UFO incidents, and of on-the-record interviews with highly placed political leaders, military authorities, and civilian pilots who discuss their firsthand knowledge of UFOs. All those interviewed have no doubt that humanity is encountering highly intelligent aliens who are technologically advanced far beyond human dreams and who are at least curiously and casually observing us, and perhaps studying us scientifically. Their visits and actions to date, for the most part, seem to indicate that they have peaceful intentions.

We'll look, then, at extraordinary, well-documented events that occurred in the U.S. and one which, while it happened in England, is directly linked to the U.S. because when it occurred the United States Air Force (USAF) was using a Royal Air Force (RAF) military base in that country.

Roswell-Corona, New Mexico, 1947

Reports in 1947 from the Roswell-Corona area of New Mexico that there had been crash landings of at least two extraterrestrial, technologically advanced spacecraft, possibly controlled by extraterrestrial beings, at least one of which was piloted or at least occupied by ETI, are well known internationally. They are among the most startling and studied UFO narratives available to date.

It's perhaps no coincidence that the Roswell event and a flurry of other UFO sightings and stories occurred in New Mexico shortly after the development and testing of the first atomic bombs, because of the surge of radioactivity that would have been sent into space. The U.S. detonated the first one in New Mexico in the Trinity test, followed shortly thereafter by U.S. attacks using atom bombs on the Japanese cities of Hiroshima and Nagasaki, after which the bombers, their crews, and spare atom bombs were located at the Roswell Army Air Force base. Thereafter, the U.S. and U.S.S.R. for decades had ongoing atomic and nuclear weapons research, development, and testing on land, in the seas, and in the atmosphere. During and after that time, as UFO overflights became more commonplace, thousands of people throughout the world reported sighting them. The Roswell crash sites, debris, and four or more ETI occupants of the craft were seen by hundreds of civilians and military personnel. The U.S. government denied that UFOs had crashed at Roswell, and used coercion and ridicule to silence those who knew otherwise.

The best of the Roswell narratives includes J. Allen Hynek's works (discussed in chapter 2); *Witness to Roswell: Unmasking the Government's Biggest Cover-Up*, by accomplished investigative writers Thomas J. Carey and Donald R. Schmitt (2009 edition); *Crash at Corona: The U.S. Military Retrieval and Cover-Up of a UFO*, by aviation/science writer Don Berliner and nuclear physicist Stanton T. Friedman (2004 edition); and *The Roswell Legacy* (2009), by Colonel Dr. Jesse Marcel Jr. and Linda Marcel, replete with photographs, which provides invaluable and focused firsthand information on Roswell. Dr. Marcel wrote the book to keep a pledge to his dying father that he would restore his father's good name and reputation, which had been impugned when the USAF claimed in 1947 that he had seen a crashed weather balloon, not a crashed UFO.

Jesse Marcel Sr. was the Army Air Force intelligence officer at Roswell Army Air Field who recovered debris from the first crash site, might have seen a second site and alien bodies, and showed debris from the "flying disk" crash to his wife and son when returning to the Roswell base late at night. A very brief summary of Roswell data and persons involved will be presented here. (*Cosmic Commons* discusses Roswell events in greater depth.)

On July 3, 1947, William "Mac" Brazel, foreman of the Foster Ranch, accompanied by a child, Dee Proctor, found debris (and possibly bodies), probably remnants of a late June or early July crash during a thunderstorm, at a site or sites on the ranch. Brazel, noting the unusual nature of the debris, the skittishness of the flock of sheep that didn't cross the debris field even to reach their watering site, and perhaps seeking a reward from the government, on July 6 reported his find and showed pieces of debris to Sheriff George Wilcox. Afterward, he had a recorded interview at Roswell radio station KGFL. Wilcox called the Roswell Army Air Field (RAAF; at that time the branch of the military known now as the U.S. Air Force had not yet been established), and relayed the information to its chief intelligence officer, Major Jesse Marcel Sr., who was in charge of the intense security of the only atomic bomb base in the world. Subsequently, Brazel, Marcel, and Captain Sheridan "Cav" Cavitt, an agent of the Counter Intelligence Corps (CIC; predecessor of the CIA), traveled to the ranch.

On July 7, 1947, Major Marcel returned to the base separately from Cavitt after their visit to the crash site, and stopped en route to show debris samples to his wife and to his son, Jesse Marcel Jr. Major Marcel reported his find and gave the debris to Colonel William Blanchard, commander of the Roswell Army Air Field (which was home to the 509th Bomb Group that had dropped atomic bombs on Hiroshima and Nagasaki in 1945, and served as a storage site for other atomic bombs).

On the morning of July 8, 1947, Blanchard announced via a press release sent by Lieutenant Walter Haut to local radio station KGFL and the local newspaper that a "flying disk" of extraterrestrial origin had been found. The *Roswell Daily Record* headline on July 8 proclaimed: "RAAF Captures Flying Saucer on Ranch in Roswell Region"; the subheadline announced: "No Details of Flying Disk Are Revealed." The news was broadcast throughout the United States. Late that same afternoon General Roger Ramey, commander of the 8th Air Force, after hearing about the news story, ordered Colonel Blanchard to issue a retraction and state that the debris found on the Foster Ranch was a crashed weather balloon. Ramey arranged for a photojournalist/reporter from the *Fort Worth Star-Telegram* to come to his office at the Fort Worth Army Air Field, where Ramey elaborated on the press release "error." Major Marcel was ordered to attend the conference without speaking, and to kneel on one knee while holding pieces of an actual weather balloon (not Roswell debris) in order that it and he might be photographed for the newspaper.

In the more than six decades since the Roswell incident, the U.S. Air Force has changed its account of what happened at least three times. The news media discovered discrepancies, each of these times, between factual data from the air force's own records regarding types of weather balloons in service when the crash occurred, and the "official stories" promoted by the air force. In 1997 the USAF announced that the case was closed, despite both lingering questions and new questions, and has not responded to more recent testimony from hundreds of witnesses to the event. Many witnesses overcame fears of retribution from government agencies and agents, and reported what they had seen themselves or heard from friends and family. Decades after Roswell, investigators interviewed Roswell-Corona, New Mexico, area ranchers, members of their families, military personnel, and members of their families. Roswell investigators have documented that military personnel were threatened with court martial, expulsion from the service, or lack of advancement if they were to reveal what they knew about Roswell; and civilians were physically threatened, and their lives and those of their families were threatened, for reasons of "national security," if they spoke publicly about what they knew about the New Mexico incidents.

Colonel Dr. Jesse Marcel, Jr., son of Colonel Jesse Marcel, Sr., is recognized today as a key witness to Roswell area events by those who accept the general narrative as historically accurate overall. Dr. Marcel (1936–2013) was a retired Helena, Montana otolaryngologist (an ear, nose, and throat spcialist, or ENT); a veteran of the first U.S.-Iraq war, in which he was a flight surgeon and a helicopter pilot who flew more than two hundred combat missions during a thirteen-month period in a 2004–5 tour of duty; and a retired colonel in the Montana Air National Guard. He continued until his death (which occurred before he could finish his reading of *Encountering ETI*, for which he had agreed to write an endorsement) to be very interested in Roswell and its aftermath. He traveled extensively and gave numerous media interviews to correct false statements about the role and message of his father, who had investigated the Roswell event as the ranking security officer of the RAAF. Dr. Marcel pledged to his father, when the latter was on his deathbed, that he would correct the misinformation distributed by U.S. government agencies and the United States Air Force about his father's professional capabilities and conduct. To that end, Marcel Jr. published *The Roswell Legacy: The Untold Story of the First Military Officer at the 1947 Crash Site*, written in collaboration with his wife, Linda Marcel; the book is dedicated to his father. Jesse Marcel Jr.'s firsthand account of Roswell events is the principal source for discussion of Roswell in this chapter.

Marcel Jr. writes that "mine is a story of actually seeing and handling artifacts from the site." He declares that the answer to the question, "Are we alone in the universe?" is "an emphatic no." He includes in the book several photos not only of the principals in the Roswell event, but also of air force documents that verify his father's expertise in intelligence matters and weather balloon characteristics and projects.

Major Marcel's responsibilities as Intelligence Officer at the RAAF included investigating "aircraft accidents, or any problem that arose with security." This was a highly important assignment, since the base had atomic weapons and was integrated within the Strategic Air Command (SAC).

Marcel Jr. describes in detail and for several pages his recollections of the night in 1947 when his father came home, briefly, carrying several boxes of Roswell crash debris (three different types of material) that he was taking to the RAAF. His father told him, years later, that he had not seen any alien bodies, but had heard reports about them. After RAAF commander Blanchard issued the "flying disk" crash press release and Ramey's press conference in Fort Worth, by July 10 all of the Roswell debris had been removed from the Foster Ranch and transported by a massive C-54 cargo plane to Wright-Patterson Air Force Base near Dayton, Ohio. Several witnesses, including the C-54 pilot, Captain "Pappy" Henderson, later reported seeing bodies carefully packed in one of the large crates that were placed on board the aircraft.

In 1978, after more than three decades of silence about what he had seen and held in his hands, Major Jesse Marcel Sr. stated to an interviewer after his retirement as Colonel Marcel (his promotions indicated respect for his professional work) and years of military-imposed silence and secrecy, that what he saw in 1947 was "not of this Earth." Immediately, public interest in Roswell was revived. Hundreds of witnesses to one or another aspect of the UFO crashes in 1947 then came forward, despite government threats and cover-ups, to provide additional information about the Roswell-Corona events.

Col. Dr. Jesse Marcel Jr. discusses government efforts to cover up the Roswell crash, and describes in detail the structure and function of a Mogul weather balloon to demonstrate that the crash debris did not resemble it in any way. He explores, too, the shutdown of the Echo Flight and Oscar Flight ICBM missiles near Malmstrom Air Force Base, Montana, in 1967.

As he ends the book, Marcel expresses the hope that readers will be "touched by the truth of otherworldly civilizations," as he was, and will "look objectively, hopefully, and perhaps even lovingly upon the promise that future contact holds for humanity." Humans should "act responsibly when faced with evidence that the universe is big enough to be home to a great number of civilizations . . . to deny this would be to place human limits upon God's infinite capacity for creation."

In a 2012 email to this writer, Jesse Marcel Jr. reflected back on his 1947 childhood experience, recalling his vivid impressions of handling the debris and his thoughts at the time. He reflected: "In 1947 existence of planetary systems was only a theoretical possibility but recent discoveries have found that planetary systems are like grains of sand on a beach. When I saw the Roswell debris, I was pretty sure that there were other earthlike planets out there populated by intelligent civilizations, so I was gratified later that telescopes like the Kepler have proven what I was thinking all along was true. Although there is no proof at present that would satisfy skeptics as to the reality of the fact that there are one or more civilizations that are aware of us and indeed have sent probes to our planet much like what we are doing in our solar system, that will come with time."

Retired Lieutenant Colonel Philip J. Corso authored *The Day After Roswell*, which focuses on decades of military research on the Roswell-Corona crashes of UFOs. What makes his narrative unique is that Corso is to date (as far as this author could find) the most senior military officer to acknowledge not only that there were bodies recovered from the Corona site, but that he had seen an alien body in temporary storage in Kansas and, years later, read a top-secret report about Roswell in the Pentagon. The documents contained a copy of the Walter Reed Hospital pathologist's autopsy of an alien cadaver from the crash site.

In June 1947, Corso was in charge of security as the post-duty officer at the Fort Bliss, Kansas, air force base (formerly a cavalry base, the home of General Custer's 7th Cavalry). He states that while he was making his rounds on the night of July 6, he saw inside a veterinary building a coffin-like wooden crate that contained, immersed in fluid, one of the alien bodies recovered from the Corona site: "a four-foot human-shaped figure with arms, bizarre-looking six-fingered hands—I didn't see a thumb—thin legs and feet, and an oversized incandescent lightbulb-shaped head... its facial features... were arranged absolutely frontally." The attached paperwork, an Army Intelligence document, described the box's contents as "an inhabitant of a craft that had crash-landed in Roswell, New Mexico, earlier that week"; the routing manifest ordered the box to be sent "to the log-in officer at the Air Materiel Command at Wright Field and from him to the Walter Reed Army Hospital morgue's pathology section." Fourteen years later Corso was stationed in the Pentagon under General Arthur Trudeau, who headed the top-secret Research and Development Division; he was assigned to the division's Foreign Technology desk. In that capacity, after being secretly assigned to explore secret Roswell files and debris to assess possible military and commercial uses of UFO technology recovered from Roswell, Corso saw "the autopsy photos and the medical-examiner sketches." He adds details to what he had described that he saw in Kansas: "The organs, bones, and skin composition are different from ours. The being's heart and lungs are bigger than a human's. The bones are thinner but seem stronger as if the atoms are aligned differently for a greater tensile strength." Corso documents government efforts to suppress and cover up all UFO reports, and to use ridicule against those who discuss them in order to discredit witnesses and dismiss incidents. The book includes UFO surveillance photos in Army Intelligence files that were used to advocate support of army research and development efforts focused on the reverse engineering of Roswell debris; appendices contain copies of previously classified military documents that analyze reports about and implications of UFO sightings, and the military's proposal, complete with drawings, for a military base on the moon, called Project Horizon, to monitor and defend against extraterrestrial spacecraft.

The "Roswell Incident" remains very much alive in public consciousness, stimulated especially by revelations from those who once had significant experiences with the event itself or with credible witnesses who were significant participants in or observers of the activities at the Foster Ranch, the Roswell Army Air Field, Wright-Patterson Air Field, or were involved in or aware of military maneuvers and policies to silence UFO discussion.

Meeting Dr. Marcel: Helena, Montana, 1989

In 1989 my wife and I took our daughter and son to the emergency room of the Helena, Montana, community hospital—both children had nosebleeds from the dry air and the city's four-thousand-foot altitude. The ENT physician on duty that night was Dr. Jesse Marcel Jr. I knew, having read about Roswell over the years, that he was the son of Major Jesse Marcel, Sr., the intelligence officer at the Roswell Army Air Field who had identified debris scattered on a nearby ranch as pieces of a crashed alien "flying disk," and showed boxes of debris to his wife and son. I had heard that Dr. Marcel was in Helena, but I did not broach the subject of Roswell with him in the hospital emergency room. A few years later his medical office manager, who lived across the street from us, learned of my UFO interest. The film *Roswell* had just been made for TV, but had not yet been released or broadcast. Dr. Marcel had received a copy of the film, for which he served as a consultant; Martin Sheen has a role in it as a CIA operative. Dr. Marcel permitted his assistant to borrow his video of the film to show to her husband and to my wife and me in her home. Once I began researching UFOs and ETI in earnest decades later when a professor at Boston University, I initiated an email correspondence with Dr. Marcel. We kept in touch periodically. I much admire his openness, courage, dedication, and persistence in promoting understanding of events that transpired at Roswell, and his father's role in them.

Hudson River Valley Region

The Hudson River runs primarily and extensively within New York State, from its source in Henderson Lake to where it divides New York City and the New Jersey Palisades just before it enters New York Harbor. UFO incidents in the Hudson River Valley were centered north of New York City, above the river or in areas contiguous to it, and extended outward to neighboring areas of the State of Connecticut.

In describing the Hudson River Valley events, Leslie Kean states that "the Hudson Valley UFOs . . . did not exhibit any aggressive or hostile behavior." Some people flashed car lights at them, and UFOs responded in kind, matching the number and duration of flashes. Kean observes, too, that in 1984 "six security guards at the Indian Point nuclear power plant witnessed the UFO hovering about 300 feet over the reactor in restricted airspace. Two guards told investigators it was a solid object bigger than a football field."

The most extensive description of unidentified aerial phenomena/on (UAP) events in the Hudson River area is presented in *Night Siege: The Hudson Valley UFO Sightings*, by J. Allen Hynek, PhD, Philip J. Imbrogno (a science educator and author of books on UAP topics), and Bob Pratt (journalist and author of UAP-focused books). The authors state in their collaborative work that there was no lack of credible witnesses to come forward and describe their experiences. In fact, at a Brewster, New York, UFO conference in 1984 and in its aftermath "so many professional people—people who normally would not speak out in public about seeing a UFO because it would damage their reputations—were now going on record as having witnessed whatever it was in the sky." They had been empowered to speak because they had seen objects whose characteristics and actions contrasted with known human science and technological capabilities; events had been publicly described in front-page articles in local newspapers; and a few professional, highly respected people working for major corporations had formed a vanguard that went public about what they had witnessed.

The authors—especially Hynek because of his previous role as air force civilian scientific consultant—are well aware of government efforts to suppress UAP information and to use false data, ridicule, and denial to coerce members of the public into being fearful of social and professional consequences should they acknowledge any experience with UFO phenomena or, in some professions, merely mention an interest in the subject. People who are credible witnesses because of their profession or educational background are especially vulnerable to reprisals, and often practice self-censorship as a result. During the research and writing of *Night Siege*, however, hundreds of witnesses came forward to speak with the authors publicly during a workshop on the subject or privately afterward. They were buttressed by hundreds of fellow witnesses and newspaper stories reporting events; through the preceding, they achieved a certain comfort level that enabled them to offer their testimony. Even so, an undercurrent of fear was present in some cases, primarily because of possible employment repercussions, to the extent that several witnesses related what they had seen but requested that their name and, for some, their occupation even without mentioning their name, not be used in the book.

In some small towns, even members of police departments were willing to go on record—again, because many of them were witnesses to the phenomena, or knew someone else who was a witness. The book includes photos taken and videos recorded by witnesses, including by an officer of the Connecticut State Police from a site near the I-84 highway. A brief summary of witnesses' stories follows.

On March 24, 1983, between 8:00 and 9:30 p.m., hundreds of people in the Hudson River Valley, most of them near Yorktown, New York, observed a low-flying, boomerang-shaped UFO with red, green, and blue lights. Although the Yorktown Police Department acknowledged receiving hundreds of calls and conducted an investigation, afterward the police told people at first that they could not provide details because of the "sensitivity of the situation," and then declared that people had seen planes in formation. Later, one of the police officers called the authors who were writing *Night Siege* and told them that he and other members of his police department did not accept the official explanation. They knew firsthand that the object was not several planes in flight: "I saw the thing, and it hovered directly over my head for five minutes—and airplanes don't hover!" The aircraft explanation—including an astounding statement that witnesses had seen ultralight aircraft flying in formation when there were very dangerous high winds—became the official FAA response to area UFO reports. There were numerous UFO occurrences for two days. Witnesses who went on record included Hunt Middleton, a New York corporate executive; Ed Burns, an IBM program manager, who said that "if there is such a thing as a flying city, this was a flying city.... It was huge"; Kevin Soravilla, a police officer; Bill Hele, chief meteorologist for the National Weather Corporation, who stated, "I have been around planes for the past twenty years—and ... I realized that this thing did not have the appearance of any known object or anything similar to an airplane or group of airplanes that I'd ever seen"; John Piccone, Grumman employee who worked with numerous types of aircraft and spacecraft; Joe Trongone, mechanical engineer, who designs small aircraft and who was sure that the UFO was not an airplane; Andy Sadoff, New Castle policewoman; Albert Silbert, a physicist and high school science teacher, and his family, who did not disclose what they had seen until months later, fearing ridicule; and Robert Golden, IBM executive. Even though government and government agency staffs continued to offer the light aircraft explanation, "a number of people who saw both the UFO and later the formation of planes said the planes looked nothing like the object." The authors note that "the caliber of people who were willing to go on record as having seen this UFO was remarkable. Many of these people were professionals who had nothing to gain and much to lose by reporting they had seen a UFO—scientists, engineers, doctors, lawyers, pilots, police officers, and many other people who are stable, respected residents of the area."

On October 28, 1983, Jim Cooke, a biomedical engineer with a background in physics and engineering (and who designs laser optics and laser surgery equipment), saw a very large and unique object in the sky. He was on Croton Falls Road, between Croton Falls and his home in Mahopac, New York, at 2:15 a.m. "It looked like no aircraft [I] ever saw," Cooke said. He stopped his car and saw it hover about fifteen feet above the Croton Falls Reservoir. Cooke provided specific details about the activities of the triangular-shaped object, which had a nonreflective surface and sent a light-like probe into the water. Cooke's sighting was unusual only in terms of his close proximity to the object and his description of its apparent scientific probe of the reservoir's water: from 1982 to 1995 there were more than seven thousand reported UFO sightings over or near the Hudson River Valley, covering an area of fourteen hundred square miles in three New York counties and three Connecticut counties.

On March 25, 1984, hundreds of people in and around Peekskill, New York, viewed a large, triangular- or boomerang-shaped UFO. On the following day the daily newspaper for a several county region in New York, the *Reporter Dispatch*, carried the headline: "The UFOs Are Back and They're Right on Schedule." Once again, the Federal Aviation Administration assured residents that they had seen stunt pilots; this angered people who had called the police department, since they knew otherwise. One witness responded, "I am an educated person. Don't you think I can tell the difference between planes and something strange in the sky?" David Boyd, a pilot who saw lights in the sky approximately eight hundred to one thousand feet overhead, thought at first that he might be seeing ultralight aircraft, but was doubtful. Then he saw the lights make a level turn (without banking). At home, he called the police and was told he had seen ultralights in formation: "I questioned that because I don't know many guys who fly ultralight in formation.... [Moreover] it was moving... much slower than anything I have ever flown. Anything flying that slow would drop to the ground."

On June 11, 1984, the *Reporter Dispatch* headline read: "UFO Buzzes New Castle." The news story elaborated: "Once again mystery invaded the night skies of Westchester and Putnam Counties. The strange object was flying and it was unidentified.... Whatever it was, it disrupted the New Castle town board work session and caused dozens of other people to look up and wonder. New Castle Police Sergeant George Lowert burst into the board room at 9:35 p.m. to announce a UFO was cruising over the building.

Within seconds, the entire board and a dozen spectators were scampering down stairs to catch a glimpse of the lights." The article reported that a Westchester County Airport control tower operator saw an object on radar at 9:30 p.m. Later, in an interview with *Night Siege* authors, Lieutenant George Lowery, who years before had been the desk sergeant described in the *Reporter Dispatch* story, elaborated further on what he and other police officers had seen. Although the FAA issued its standard "airplane" line, Hynek et al. observe, "A police officer wouldn't interrupt a public meeting of the town board just to have the officials run outside to see some planes."

Sightings on June 22 were videotaped by Joe Bova and his family. The authors took the tape to a "highly trained pilot with thousands of hours of flying time in all kinds of aircraft," who "has a background in physics and electronics" and has worked on guidance systems. A consultant to the U.S. government, he declined to be identified by name. After viewing the tape for three hours on his sophisticated equipment, he said, "No, not planes," and went on to explain his reasons: the lights stayed fixed in relation to each other during changes in direction, which would not occur if several planes were flying in formation; the lights were too bright and the motion too stable for small aircraft; and the type of "formation" that would have been required would not be used even by the best military pilots flying the best military aircraft, because it would have been too dangerous.

On July 12, the largest number of sightings to date was reported. Witnesses included Ed Mulholland, engineer for the Perkin-Elmer Corporation, which "designed and built the optics and some of the electrical components for the NASA Hubble Space Telescope"; Dr. Richard Long, university law professor; a minimum of twelve Danbury police officers, including Lieutenant Kevin Barry and Chief Nelson Macedo, who was fishing with his brother-in-law, Charley, and a nephew when they saw the UFO, which had twenty to thirty red, orange, green, and blue lights—when Charley shut off the boat's lights, the UFO did the same with its lights, and when Charley flicked the lights on and off, so did the UFO; a Bethel police officer who trained his spotlight on the UFO, which then shined a brilliant light over him and people around him; and Mark Purdy, construction engineer. On July 24 Bob Pozzuoli, a major electronics firm vice-president, videotaped a UFO. An ABC technician reviewed the tape and said, "I've never seen anything like it before. Every time I watch it, I get goose bumps. It's weird." Next, Dr. John Baker, of West Coast University, California, analyzed it, said it was not an aircraft, and withdrew from the analysis, perhaps because of concern about professional or personal repercussions, or his relationship with a related government entity. In 1985 HBO wanted to use the tape for a television series, and sent it to Dr. Al Hibbs at the Jet Propulsion Laboratory, Pasadena, California. Hibbs used a highly sophisticated computer that "was used for imaging data from the Voyager and Viking space probes, which sent back pictures of Mars, Jupiter, Saturn, and Uranus. Hibbs concluded that the object in the Pozzuoli tape could not be identified," and this continues to be its status. Also on July 12, Bill Sockey, "who has a master's degree in the philosophy of science" and is an officer in a Catholic Church lay association, saw an "arch of white lights" when he and his wife were returning home after a "course at the diocesan seminary." He said it was either stationary or moving very slowly. When asked whether it made noise, he responded, "It was silent. There are things in my background that I think are significant to the sighting. Specifically, I spent four years with the navy. I was a naval officer in Vietnam. I was trained in observing aircraft. . . . I was a ship's navigator for one year. So I was trained in constantly, morning and evening, observing the sky, constantly seeing the stars and the planets. I could pick planets out simply because I saw a star with the other positions in the heavens."

On July 24, 1984, the same evening on which Bob Pozzuoli shot his video in Brewster, twenty miles away, a major close encounter occurred above the Indian Point nuclear reactor on the Hudson River. After the event, a New York State Power Authority police officer on security duty at the reactor complex called the authors to say that he and eleven other officers saw the UFO when on duty. A "giant UFO" had hovered three hundred feet above a reactor "for more than ten minutes." The guard said that the shift commander "was ready to order the guards to shoot the UFO down." When interviewed by the authors, with the reluctant permission of Indian Point administrators and under the watchful eyes of a security supervisor, "Carl" (a pseudonym) said that he and others, including Con Edison utility personnel, had seen a UFO on both June 14 and July 24. On June 14, the UFO appeared to have a boomerang shape. He estimated that it was about three hundred feet in length and moved at approximately ten miles per hour. He declared that "no small planes could stay in formation with the wind that night. The wind didn't faze these lights at all. When it hovered, it just stood there. I was in the service and I flew helicopters, and I know how hard it is to keep formation with small planes. . . . The lights were much too intense for a small aircraft." On July 24, a guard yelled, "Hey, here comes that UFO again!" Two supervisors and three officers ran out to observe. It hovered only thirty feet above the reactor. A camera on a ninety-five-foot pole was directed to videotape the UFO, which was "bigger than a football field," for some fifteen minutes. Although this and other cameras usually constantly record on video tape all objects near the reactor, authorities said that no footage exists of the event.

Close Encounters? Poughkeepsie, New York, 1963

On a clear, starry night in 1963 I sat on the west bank of the Hudson River with two college friends, absorbed by the stellar beauty of the sky. While we sat in silence, suddenly I saw an extremely bright meteor arc across the heavens from south to north, above the river. I watched it, entranced by its size, speed, and bright light as I waited for it to burn up in the atmosphere and disappear. It didn't. Instead, I saw the impossible happen: Without stopping, and without any banking curve, it abruptly went perpendicular to its original course without diminishing its velocity, and after a few seconds disappeared into the night sky. I blinked in amazement, and thought, "That's impossible. I've studied physics. Anything going that fast would have shattered had it suddenly changed direction like that. I must be mistaken." I remained silent. One of my friends spoke: "Did you see that light go at a right angle to itself?" Somewhat relieved, I replied that I had. My friend observed, "I guess we just saw a UFO." We laughed, and agreed that it was so.

A quarter-century later, I stood by the cash register near the front of a mall bookstore in Helena, Montana. I was waiting for my ten-year-old son, who was getting a book from the children's section in the back. I glanced at a pile of books on the counter by the register. The subtitle of one prompted me to pick it up and page through it: UFO Sightings in the Hudson River Valley, 1962–1965. I remember that the book stated that there were several hundred UFOs seen in the years around the time of my own "meteor" experience. Regrettably, I did not buy a copy. I thought, "I was there. Interesting."

Lying Eyes? Boston, Massachusetts, 2011

At a faculty meeting at Boston University I discussed my current research and writing. I described my particular interest in events at Roswell and in the Hudson River Valley. Afterward, a colleague told me that he hadn't heard "that" in a long time. "Roswell?" I queried. "The Hudson River UFOs," he replied. It turned out that his father had seen the UFO(s) flying over his home in the Hudson River Valley. Soon, U.S. government officials, propaganda units, and news control agents held meetings and workshops at which they proclaimed that the lights that my friend's father and other credible witnesses had seen were ultralight planes flying in tight formation—even though it was a very windy night. My colleague said that his father would go to such meetings and contradict the government officials, asking probing questions. Until his dying day, he said that he'd seen a UFO, and would ask at meetings and among family and friends, "Who am I supposed to believe: my 'lying' eyes or my government?"

Rendlesham Forest

The British government, responding in 2008 to a Freedom of Information Act request, released extensive documents from the Ministry of Defence (MoD) about reports of UFOs landing on British soil. The British Broadcasting Corporation (BBC) examined these files, and others released in subsequent years. The previously secret files discuss MoD efforts to track and document appearances of UFOs as reported by British citizens, military personnel, and at least two United States Air Force officers—one from his fighter plane, and another at a base in England used by the USAF—from 1978 to 1987. The BBC perused the files and did two reports on Rendlesham Forest events in 2008, one published (and at one time available online) and the other an online audio tape of a televised BBC News broadcast.

The published report, posted on the BBC News website on May 13, 2008, appears to be unavailable now. The print version, still available, has this headline: "Secret files on UFO sightings have been made available for the first time by the Ministry of Defence."

The article as originally printed states that the UFO files "include accounts of strange lights in the sky and unexplained objects being spotted by the public, armed forces and police officers." One file states that "the United States Air Force filed a report about two USAF policemen who saw 'unusual lights outside the back gate at RAF Woodbridge'" in Suffolk, in 1980. The latter event gained increasing notoriety as the "Rendlesham Forest Incident," and will be discussed below.

In 2010 a headline announced, "Churchill ordered UFO cover-up, National Archives show." The narrative relates "an account of a wartime meeting attended by Winston Churchill in which, it is claimed, the prime minister was so concerned about a reported encounter between a UFO and RAF bombers [that were returning from a WWII bombing mission], that he ordered it be kept secret for at least 50 years to prevent 'mass panic.'" Perhaps Churchill had in mind the "mass hysteria" that occurred in the U.S. on October 30, 1938. As noted earlier, in the early evening on that date, on the CBS Radio Mercury Theatre program, Orson Welles directed and narrated an adaptation of *The War of the Worlds*, an H. G. Wells novel (written in 1898) about a Martian invasion of Earth. People across the U.S. thought it was a series of actual news reports by frightened and frantic reporters in the field, whose urgent bulletins kept interrupting an originally scheduled music program (Welles narrated the story interspersed with music, as planned), and became afraid that an actual Martian invasion was underway. Government offices and newspapers were inundated with phone calls from alarmed citizens; some of the callers stated that they had seen alien craft in the skies. Apparently, some U.S. government officials and high-ranking military officers had concerns about a similar "mass panic" when they saw Colonel Blanchard's Roswell press release, almost a decade later, that a "flying disk" had crashed in the region where the Roswell Army Air Field base was located. For that and other reasons (perhaps to try to "reverse engineer" alien debris and convert it to U.S. military and industrial purposes), they decided to cover up the event and all traces that it had occurred—including through the intimidation of credible witnesses.

In 2012, the Ministry of Defence released almost seven thousand documents covering the period from 1996 to 2008.

The most extraordinary and well-known UAP event in the United Kingdom occurred in England in December 1980, in the Rendlesham Forest area near RAF bases being used by the USAF as part of NATO joint operations. The events described came to be known collectively as the "Rendlesham Incident." BBC Newsnight's "Today programme" audio file was placed online on May 14, 2008, with the headline "UFO sightings files released." It includes the actual recording of USAF Colonel Charles I. Halt's live description of events transpiring at Rendlesham Forest as he observes a spacecraft doing maneuvers in the sky. In the first part of the recording Colonel Halt, who was Deputy Commander of the USAF Bentwaters base, East Anglia, sees something in the sky, observes "We've got an object," and describes the distinct directions in which it flies. In the second part, recorded two days later, Halt exclaims excitedly—and, it seems, nervously—"Here he comes from the side. He's coming toward us now. Now we're seeing what appears to be a beam coming down to the ground. This is unreal!"

Leslie Kean provides extensive details about the Rendlesham Forest event that is described in BBC reports. She examined reports and interviews provided by U.S. military personnel who are retired now but were present at the base at the time of the incident. She interviewed two of the base personnel, Sergeant James Penniston and Colonel Halt. Besides them, numerous members of the U.S. military witnessed the craft both in the air and on the ground; all were stationed at NATO's joint U.S.-U.K. bases: RAF Bentwaters, East Anglia, and RAF Woodbridge, Suffolk.

During and immediately following his experience, Sergeant Penniston, who was one member of a three-member security police squadron sent out to Rendlesham Forest to investigate the object that had been observed landing, described in his logbook what he saw. He drew pictures that depicted a triangle-shaped craft with symbols on its side. He touched the craft when it was on the ground: it was warm and metallic, but "smooth as glass." After forty-five minutes, it lifted from the ground without any air disturbance noise; subsequently, it "maneuvered through the trees and shot off at an unbelievable rate of speed. It was gone in the blink of an eye. In my logbook, which I still have, I wrote 'Speed Impossible' . . . I knew that this craft's technology was far, far above what we could ever engineer. . . . I realized that I was 100 percent certain that we are part of a larger community beyond the confines of our planet." Kean's book *UFOs* contains Penniston's drawings, copied from his notebook.

For his part, Lieutenant Colonel Halt went out to the site when he was told that the craft, which had left, had returned. He intended to "debunk" the incident. He had a handheld tape recorder on which he recorded his foray with his patrol, as described above. There is major tension in the team's voices as a hovering craft with bright lights descends, and flashes lights around them. Halt's subsequent "Unexplained Lights" report observed that the UFO was "metallic in appearance and triangular in shape . . . a pulsing red light on top and a bank of blue lights underneath. . . . The animals on a nearby farm went into a frenzy." Later, an investigative team from the base found that the residual radiation on site was very high: about seven times higher than the ordinary background radiation in that area. RAF radar operators had been tracking the object simultaneously, and stated that they saw something "cross their screen at extremely high speed, up to 3,000 to 4,000 mph."

According to Leslie Kean, years after the Rendlesham Incident, five-star admiral Lord Hill-Norton (the UK's former chief of the defence staff, equivalent to chairman of the joint chiefs of staff in the U.S.) became "particularly outspoken on the Rendlesham Forest case and felt strongly that the MoD's line on the incident (that the events were of 'no defense significance') was entirely unacceptable and at odds with the facts." Hill-Norton, according to Georgina Bruni in her book *You Can't Tell the People*, sent his letter to Lord Gilbert at the House of Lords. It stated the following:

> My position both privately and publicly expressed over the last dozen years or more, is that there are only two possibilities, either: *a*. An intrusion into our Air Space and a landing by unidentified craft took place at Rendlesham, as described. Or: *b*. The Deputy Commander of an operational, nuclear armed, U.S. Air Force Base in England, and a large number of his enlisted men, were either hallucinating or lying.
>
> Either of these simply must be "of interest to the Ministry of Defence," which has been repeatedly denied, in precisely those terms.

The Rendlesham events continue to be a source of interest and debate in the U.K. and beyond. The "missing" files still have not been released to the press and public. (Given that the British released most of their own reports and that the Rendlesham events involved U.S. military personnel, it is possible that the files "disappeared" because of U.S. pressure on, or a "friendly request" to, the British government.)

"Where's the Evidence? I'm from Missouri, the 'Show Me' State"

Some people refuse to accept evidence of UFOs or ETI even in the face of credible witness accounts, astronomical discoveries, mathematical probabilities, and confirmed scientific data. They accept the U.S. government's absolute statements and propaganda from nonscientists that UFOs do not exist—let alone ETI who are controlling, from a distance or as a crew on board, unidentified space vehicles. (They might also be afraid, at least subconsciously if not consciously, of the third displacement discussed above.)

Why has there been no sustained presence of ETI, some dramatic, extended hovering over a large urban area? Shift your perspective from that of the representative human who experienced the first and second displacements, to what view you might have if you were an explorer of a highly intelligent alien species that observed human conduct on Earth. You would realize that technologically you are significantly superior to the dominant species you see. You would note how they treat each other, including in wars, by food and water deprivation, and through political and economic domination that is militarily enforced or economically coerced. You would be horrified to see that they are extincting other biota, and generating types of air, land, and water pollution that threatened to destroy their only home. As you record and reflect on your observations, would you want to be in Contact with such a species or even select members thereof? You'd be even more cautious, of course, if their military airplanes aggressively assaulted you with missiles and cannon fire in their planet's skies—which happened above Earth on several occasions, as attested by air force pilots from several countries. You'd likely not dare to try to make Contact with the opening words, "We come in peace"—on your highly visible descent to Earth you might be aerially assaulted before you even came close enough to communicate with Earth citizens. So, surreptitious trips to Earth, and Contact with a few selected Earthlings, would be the intelligent course to take—as ETI apparently has done.

Retired USAF Captain Robert Salas had Contact with UFOs when one shut down his underground control center for his flight of ten ICBM missiles in separate underground silos near Great Falls, Montana, in 1967. In reflecting back on this and other ETI events throughout the world, Salas wrote in a 2012 email to this writer (discussed in *Cosmic Commons*) his assessment of his experience and the reason why ETI has not made sustained Contact: "When you get past the question of the reality of this phenomenon, you search for the answer to the question, why are they here? Clearly they have announced their presence to all humanity. . . . I think the main message is that we humans seem to be bent on our own destruction by the way we are treating ourselves and our home planet. Nuclear energy and weapons is just one aspect of this and, because of my experience, that has been my main message whenever I speak or write. . . . We can only resolve these questions of the environment and our proclivity to be warlike through cooperative actions. . . . I think ET wants to see some results in this direction before making full contact."

Salas reiterated his concerns in a 2013 email to me: "My incident involved ETI encounters with nuclear weapons. I have made it a passion of mine to promote the abolition of nuclear weapons from the point of view that this is also an admonishment by our ETI visitors in order to extend the life of our civilization by the avoidance of nuclear war, and as to how nuclear war could affect not only our planet but the Cosmos.... The incidents that I describe in my books are of major significance because of the credibility of the witnesses, extent of the evidence, and the clear and continuing message of ETI."

Salas and other credible witnesses to UFO phenomena think along the lines suggested earlier: we humans must become mature before ETI will risk not only Contact but long-term collaboration in a cosmic community.

The Three *R*s: Cosmic Displacement or Cosmic Distribution?

Events in the regions around Roswell, the Hudson River Valley, and Rendlesham might be denied because of true skepticism; subordination of one's own thinking to U.S. government propaganda, even when it is presented by government spokespersons with no scientific background; direct academic or government coercion; fear of ridicule, even from family, friends, or colleagues; fear of impacts on strongly embraced theist or atheist beliefs; or fear of psychic or physical cosmic or planetary displacement. These attitudes become increasingly tenuous since claims of Contact not only linger from the past and are cited in the present, but continue to come from around the world today and are anticipated to occur in the future.

As a concluding (for now) thought, let us ponder not cosmic *displacement,* but cosmic *distribution* as a possible or even likely consequence of Contact. In this scenario, humankind would not be displaced or extincted by a malevolent, bellicose, invading ETI (as seems to be most popular among movie producers and television advertisers; sales and profits are determining factors, of course). Rather, ETI would wait for humankind to progress in its political, economic, and spiritual evolution, in hopes that we mature before we destroy ourselves. Roswell, the Hudson River, and Rendlesham events might signal an ETI interest in Contact, and an indication that we would be considered ready for Contact when and if we were to reach species maturity. This evolutionary progress would be indicated by our just, compassionate, and environmentally responsible species consciousness and conduct not only within particular geopolitical entities (nations) but in a community of human communities, and extended outward toward other biota and our Earth home.

Our future prospect, then, would not be that we'd be displaced by ETI in whole or in part from our Earth places, but that we'd learn to develop a responsible distribution of our species, as TI-become-ETI, in other worlds in the vast cosmos, in a congenial relationship with ETI who already have been exploring the cosmos. Perhaps there already is a confederation of benevolent intelligent beings that would welcome us as a part of their common efforts to do good and live well wherever they travel. They might be related, as distinct intelligent species from different worlds, within collaborative communities that integrate all mature intelligent beings. In this cosmic community, they would seek not to satisfy narrow self-interest, but to promote cooperatively their own and others' wellbeing in a cosmic commonweal. In a future time, when we've recognized and responded to our responsibilities to all members of our own species, to other biota, and to Earth, our common ground, we, too, might be part of such a confederation.

4
XENOPHILIA AND XENOPHOBIA

Good Alien or Evil Alien from Space?

One of the worst fears that people have about Contact with ETI is economic. They are concerned that, just as segments of humanity have done throughout history, ETI will try to acquire, by whatever means necessary, humans' territory and natural goods. In North America this would mean that ETI would act with a Discovery Doctrine ideology analogous to that operative for European explorers and colonizers when they encountered indigenous populations. ETI might similarly displace humankind around the globe or in geophysical segments of Earth in order to take humans' territory and the goods which have sustained TI individuals and communities. Will "what goes around comes around" occur here?

In the now classic *Star Trek* television series, the two major and incompatible possibilities of ETI conduct—benevolent or malevolent aliens—were embodied in Captain Kirk of the starship *Enterprise* as it explored space, and the warrior Klingons who voyaged not to satisfy their curiosity about the vast cosmos but to engage in conquest missions to add to their imperial holdings.

Kirk versus Klingons stereotypically represents ancient understandings of conflict in the heavens between angelic and demonic beings, which are replicated on Earth in humans who represent and are affiliated with one or another divine being. In *Star Trek* the forces of good (benevolent humans on the *Enterprise*) are locked in battle with the forces of evil (belligerent Klingons). Since humans imagined and concretized such future conflict, we know well what the outcome will be. However, in real-world (better yet: real worlds) contexts, TI-ETI conflicts, should they occur, will probably not be so predictable.

Interestingly, there's an unintended parallel in traditional Jewish religious thought, which is firmly grounded on Earth but speculative about exoEarth beings and behavior. A Jewish friend told me that in the Jewish tradition the only "extraterrestrial" creatures, so to speak, are angels and demons. (The Most Holy, YHWH, G-d, is not creature but Creator, and creation is external or *extra* to the eternal Creator.) Intelligent beings who are approaching and arriving on Earth are *extra* to Earth, thereby extraterrestrial material beings. Humans approaching and arriving on planets other than Earth—exoEarth worlds—are ETI in those places and times. They and we, in cosmos locales and moments, are aliens to current inhabitants.

Among all these *extra* creatures probably ETI, like TI, have evolved to a biotic stage that enables them to come to know not only their own complexity but also that of their world (geophysically, ecologically, chemically, physically, and biologically). Intelligent biota (and less complex biota, too) affirm their species-oriented self-interested perspective, and alter the status of other biota and even abiota. In their view, others' inherent, intrinsic value in and for themselves is displaced by an extrinsic, instrumental value for others seeking to meet their needs and wants. Intelligent biota also develop tools to help them in this effort, which might include predation on biotic prey, or altering environments to provide shelter (prairie dogs burrow in the earth, crows nests in trees, beavers build dams, for example). When communal life and consciousness, community concern and compassion, and ecological respect guide intelligent biota, they do not oppress members of their own species, and are respectful of the intrinsic value and natural rights of territorially situated fauna, flora, and terra and use them instrumentally only as needed.

Science fiction novelists and screenwriters, and members of the general public who ponder such things, have portrayed contrasting types of aliens that might make Contact with humanity: *benevolent* beings who will provide life-enhancing technology, and persuade or compel people to live more justly and responsibly; and *malevolent* beings who will employ death-dealing technology to eliminate human competitors for the natural goods the aliens want to exploit, despite any social and ecological consequences.

If the Roswell and Hudson narratives are historically accurate, then Stephen Hawking's pessimism about intelligent life in general, including humankind and alien intelligent beings, is misplaced: it should apply solely to humans on Earth and in space, since humans are the intelligent species from whom he extrapolates his theory about ETI consciousness and conduct. Should this be the case, then ETI craft hovering over the Hudson (including over the Indian Point nuclear power plant) and UFO crashes at Roswell-Corona were possibly the result of initial exploratory ventures, undertaken without intending to make Contact or to interfere with human uses of lands, natural goods, and biota or even to acquire natural goods not being used by humans. Some people share Hawking's fears and projections, including many of those who claim to have been abducted by aliens. Perhaps two types of apparent ETI presences are indicated by two much-debated incidents: the cattle mutilations done with exceedingly precise surgical instruments, leaving no blood in the animals or in the area; and the perhaps playfully done crop circles and intricate geometric patterns carved into English fields for almost five centuries.

Aliens' Possible Responses to Roswell Events

Thinking "as if" Roswell narratives describe an actual event, a pertinent question is this: if "unofficial" witness accounts of events are accurate, including testimony that aliens were aboard one or two of the crashed objects and one alien survived but died in custody, then why haven't the aliens whose associates crashed at Roswell sought to recover their remains, rescue the survivor, or retaliate for humans' imprisonment of the survivor? The aliens are significantly more technologically advanced than we are, given the materials used in their craft, the maneuverability of their vehicles, and their ability to traverse vast distances; surely they would be more militarily competent as well.

In response to the question, we'll consider eight possibilities, among many that might be suggested. (Readers might speculate about—or know?—others.)

The first possibility is that there are, as with human groups and individuals, diverse and even conflictive cultures and subcultures in alien worlds, on the same planet, on distinct space vehicles, or on celestial bodies located at great distances from each other. One analogous expression of conflictive diversity may be found in Wendell Berry's book *The Unsettling of America* in which, relative to humans involved in agriculture who either exploit or engage with their environment, he uses the terms "exploiters" and "nurturers." *Exploiters* focus on efficiency; profitability; how quickly and how much the land they control as owners or managers can produce; and serving an organization or institution. *Nurturers* focus on care; health for land, family, community, and country; the land's carrying capacity in itself, not its desired productivity imposed externally; and serving land, home, community, and place. We can adapt these concepts of exploiter and nurturer to TI-ETI interaction. In this light, we can consider possible attitudes that ETI aliens, like ourselves, might have regarding new people or places: to interact positively and constructively, or negatively and destructively; to build up or to destroy resident aliens' cultures and peoples; and to develop or to devastate planetary environments, including biotic communities, without regard for evolutionary processes or bioregional requirements. In other words, on their home planet and other worlds intelligent beings can adapt to a place, or adapt a place to themselves; alter a place to fit their apparent needs and their desires or alter themselves to fit into a place; find their niche and become integrated with other biota and their common abiotic place, or exploit not only a place but local biota—including other intelligent beings—and compel them to assist dominant intelligent beings to extract goods far beyond what the niche can sustainably produce ecologically, culturally, and bioregionally. It is possible that in the Roswell event ETI aliens were in Berry's "nurturer" category, and reasoned that the ordinarily belligerent humans were a primitive culture, adolescents in their evolution to full humanity and acting like it, and less thoughtful about or responsible for their actions; humans' fear of the "alien" might have been acknowledged also. If ETI was in Berry's "exploiter" category, it is possible that they decided, for whatever reason, that the time was not appropriate to aggressively attack to retaliate against humans (perhaps because of insufficient vehicles and personnel on hand), and they would choose a better time. Finally, if both types do exist among ETI species and cultures, there is probably some sort of balance maintained, perhaps by a recognized intergalactic security force, similar to Earth's U.N. peacekeepers, that ensures they do not conflict—and even, perhaps, that the exploiters do not interfere with primitive cultures at whatever their stage.

Second, possibly as a consequence of a division such as just elaborated, ETI might have used its robotics technology to engineer artificial intelligence (AI) to have a high level of competence (as in the film *Blade Runner* [1982], in which highly intelligent humanoid robots resist being eliminated by humans); to exhibit "programmed emotion" (as in the film *A.I.: Artificial*

Intelligence [2001], in which a boy named David is programmed to love humans, to provide for an emotional need that a person, couple, or family might have); or to fly space vehicles or exploratory craft, including, at times, as vast storehouses of knowledge (such as the android Lieutenant Commander Data in *Star Trek: The Next Generation*, and David in the film *Prometheus* [2012]—both of whom aspire for personhood or humanness to some extent). In such cases, ultimately A.I. is not valued intrinsically, only instrumentally, and therefore is ever regarded by humans as disposable. The poignant ponderings of advanced levels of A.I. is presented in several films, where very intelligent engineered "humans" with computer brains, in appearance indistinguishable from biological humans, question why they are disposable: effectively, why they must be "terminated" or "die" at some point. Whether human engineering could ever get to the point where it could manufacture robots that reflect on metaphysical questions is much debated in the scientific community (perhaps more than in the sci-fi community). In any case, the human-appearing robots are *aliens* to us from the moment they are produced and can act individually until they are terminated: they lack our DNA and evolutionary history, and are not physiologically biota, let alone members of our species. Despite their outward appearance, extraordinary intelligence, and complex capabilities, they are not biological descendants of *homo sapiens*. Similarly, ETI might have engineered personnel in their likeness to operate some of their spacecraft or exploratory vehicles launched from them.

Third, ETI might have genetically engineered, among its own progeny, distinct classes of species members to serve different purposes, and cloned them over generations. An Earth novel that focuses on precisely this possibility is Aldous Huxley's classic narrative of a socially stratified dystopia, *Brave New World* (1932). In it, humans are developed in test tubes in a regional Central Hatchery. The unified planet's motto summarizes social roles linked to social rigidity: "Community, Identity, Stability." People's individual and group characteristics and capabilities are determined *in vitro* by the amounts and types of nutrient liquids they receive, whether or not they are given selected additives of harmful liquids to retard their growth mentally and physically, their extent of exposure to heat and cold, etc.; and, after they are born, Pavlovian practices teach them to like or dislike such things as flowers and books. Consequently, they range in "native" intelligence from Alpha Plus, through Alpha, Beta, Delta, Epsilon, and Gamma, with distinct abilities and assignments genetically induced in all cases.

Each category has its plus and minus subcategories, for example A+ (Alpha Plus) and B- (Beta Minus). Each group and subgroup is preprogrammed to fulfill specific tasks to maintain social stability, smooth institutional and mechanical structural constancy, and people's general physical wellbeing. There's a place for everyone, and everyone is in their place—irrevocably.

Fourth, alienkind chooses to exercise a policy of noninterference as their response to encounters or unintentional Contact with other intelligent beings. In this scenario, ETI will do its utmost to avoid Contact. This particular policy, noted frequently in science fiction, is highlighted currently in the latest installment of the *Star Trek* series. In the movie *Star Trek: Into Darkness* (2013), the *Enterprise* crew faces a dilemma: should they obey the noninterference order and allow a comrade to perish, or should they rescue their friend even if it necessitates disobeying the order?

Fifth, ETI is not vengeful but might, in circumstances such as an attack on their own world and community or against their spacecraft, respond defensively in kind. In Roswell, ETI who knew that some of their own were killed and that one of their own (perhaps hurt from the crash) had been taken captive, might have viewed humans as too "primitive" to have developed or evolved to a state of consciousness and a manner of conduct that is sufficiently "civilized," and humans' actions from ignorance are judged to be an unfortunate indication that humans have an arrested social and psychological evolutionary development.

Sixth, ETI regards the crashes as accidental: they occurred in a thunderstorm and their vehicles might have been struck by lightning or experienced some weather-induced magnetic interference with their computers, or electrical bursts that rendered their equipment dysfunctional, causing the craft to be uncontrollable and to crash. Humans' recovery of alien bodies and apprehension of an alien crew member might have been viewed as all part of a single "accident event" and, moreover, any attempt to rescue a survivor would have required significant Contact, which they wanted to avoid.

Seventh, ETI might have given the "benefit of doubt" to humans, and decided that human scientists and surgeons and accompanying medical personnel had rescued and were trying to save their kin; had to understand first their kin's physiology in order to do so; and performed medical experiments to accomplish those ends.

Eighth—the most intriguing and perhaps least likely possibility—ETI *did* respond discreetly and secretly in some way soon after Roswell-area events. The USAF and ETI made Contact shortly after the Roswell crash, and came to some agreement about the disposal or use of debris recovered, the disposition of the alien remains taken from the site, and the return of the surviving alien to their spacecraft. A *quid pro quo* might have been some form of information exchange and technology transfer. ETI, in this case, might have begun to communicate with the top-secret cadre in the CIA and USAF that deals with UFO/ETI events. Some diplomatic and perhaps economic arrangements (such as some technology transfer in exchange for a U.S. noninterference and denial policy regarding UFOs) might have been established. This would correspond to or at least be related to foreign governments' suspicions that the U.S. is keeping detailed records of and evaluating UFO reports from around the world—all the while claiming disinterest. This might be a source, too, of rumors and stories about so-called men in black who visit sites where witnesses have reported UFOs and who have been seen, it is claimed, in urban areas too. (Of course, the "men in black" sightings could be hoaxes.)

Possibility eight seems unlikely, particularly in light of Pentagon Col. Corso's claim that he saw an alien corpse in Kansas preserved in a crate going to Bethesda Naval Hospital and, decades later, while working on secret projects in the Pentagon to reverse engineer Roswell debris, also saw the Bethesda pathologist's report on the alien body. The latter claim itself has some intriguing possibilities: that the U.S. government used Corso to disseminate false information about an alien body precisely because his story would throw off media reps trying to find the truth about Roswell events, which really did include post-crash Contact; and, that the USAF/U.S. government was literally speaking the truth when it claimed that "there's no such thing as UFOs": precisely because the U.S. has identified positively, including perhaps through Contact, what the UFOs are—they are no longer "unidentified"—and who controls them from within or afar. Possibility eight seems the least likely of the possibilities—all of which, of course, are just speculation—but each could reasonably be considered, no matter how "far out" it appears at first thought. The former Canadian Minister of National Defence stated, in a 2013 email message to me, "Certainly the United States government has been working with at least two species and probably more. My guess is that there have been many others who have had face-to-face contact with U.S. and other countries' officials. There have been many demands for greater government transparency in respect to the ET presence and technology. It is just possible that if the U.S. government doesn't come clean within a reasonable length of time that Russia or one of the other powers will."

An alternative to speculation about direct intentional Contact is that since the U.S. has had decades to examine the debris and reverse engineer materials or even technology taken from the crashed craft (and from others at other U.S. sites), they have identified what the objects are: alien craft from somewhere in space. While—if—ETI was considering any or several of the preceding possibilities about ways to respond, they would have noted human government and military efforts to completely clear away all traces of ETI's accidental arrival. They might have wondered if TI hoped that the vehicles were unaccompanied by others, that the personnel that staffed or guided them were robots or humanoids, that somehow ETI would not be aware that the spacecraft had crashed on Earth or on human territory, or that remains of ETI occupants and remnants of ETI technology had been recovered.

The United Nations Outer Space Treaty (OST), to which the U.S. was a State Signatory, required that a nation that recovered space debris launched and owned by another Signatory had to return the debris to its owner. The U.S. might have decided (if it even considered OST requirements) that the provision did not apply because the object from which the debris was recovered did not belong to any Earth nation. Or, along the lines of ETI response possibility eight: the debris or parts thereof have been returned to the ones who launched it.

Both/And?

While Stephen Hawking is decidedly pessimistic about extraterrestrial intelligent beings' character and motives, other scientists are speculatively optimistic. The assessments of Robert Wright and Peter Singer differ from Hawking's assessment on scientific grounds; both Wright and Singer base their conclusions on species' evolution to being complex and intelligent, and living in socially cohesive or at least integrated settings. Wright and Singer theorized that ETI would more likely be benevolent, neutral, or at least nonaggressive toward other intelligent beings, their places, and the other biota with whom humans were interrelated.

The possibility that ETI is benevolent or at least neutral when it comes into Contact with new worlds and a new species might not seem very far-fetched when the complementary claims made by Robert Wright and Peter Singer are considered. In his *New York Times* opinion piece "Ethics for Extraterrestrials," Wright suggests two reasons why ETI might be benevolent rather than malevolent. First, he describes Peter Singer's view that historically humans have progressed greatly in social systems formulation and social conduct, as evidenced by changes made from the time of the Greek city-states, in which citizens in opposing states were regarded as subhuman (and as evidenced, too, in Spanish explorers' efforts to designate indigenous peoples as "beasts who talked" in order to enslave them, a historical practice not mentioned by Wright), to the twentieth century when people of diverse races, colors, and creeds are regarded as human. Singer believes this indicates a natural process in the evolution of intelligent beings, who eventually use their reason and interaction to discern that moral concern is socially beneficial, and even will extend such concern to all sentient being. Second, Wright expresses his own, contrasting perception that "pragmatic self-interest," not reason, promotes moral progress. Integrating both perspectives, Wright concludes that humankind will need such moral enlightenment—however humans evolved—to progress sufficiently—technologically and socially—to journey from its solar system farther out into space. He observes that what he and Singer theorize to be necessary for human development can be extrapolated to ETI: for them to have developed the technology to venture vast distances, they must have evolved socially along these lines. In a hopeful contrast to Hawking, the Wright and Singer hypotheses anticipate that ETI would not be the menacing, conquering species that people on Earth would not want to encounter. Both see intelligent species' evolution as the reason why benevolence would characterize them, Singer suggesting that intelligent species rationally conclude that ethical sociality will benefit them as a species (and therefore would act similarly when meeting other intelligent species), and Wright stating that moral evolution enables this to be the case. Therefore, from both scientists' thinking it might be extrapolated that humankind should be optimistic that ETI would not be aggressive and bellicose toward intelligent beings on other worlds; rather, ETI would be accommodating to and even benevolent toward them, and promote thereby a cosmos-progressive interspecies harmony.

It is possible that Hawking, Wright, and Singer are all correct. In this case, intelligent species might have evolved biologically and culturally in diverse ways on different worlds in the cosmos. On one planet (or many planets), much as on Earth, there were competing nationalisms or other

Xenophilia and Xenophobia

cultural distinctions. The most warlike, aggressive, technologically developed, and ruthless ETI triumphed over its competitors (much as some nationalistic politicians and people hope to do on Earth), and eliminated or subordinated them to its will and purposes. On another planet (or other planets), much as on Earth, competing cultures clashed, but with far different consequences and resolution. In this scenario, rational assessments, and respect for others (or, the realization that, as happened later in time on Earth, they were heading madly toward mutually assured destruction) led them to strive successfully to subordinate "-isms" to mutual wellbeing; perhaps a United Nations-like organization was established, and became respected and functional—unlike on Earth where, unfortunately, the more powerful nations on a regular basis suppress the legitimate aspirations of the less powerful nations, despite lofty U.N. ideas and ideals expressed in treaties, conventions, covenants, and international law. The success of ETI's efforts enabled them to resolve differences, share technology, establish global rights and responsibilities for individuals and political entities (equivalents of nations on Earth), restore and conserve their planetary common ground and their species common good and, as a result of all this, collaboratively develop and conjointly operate the technology and vehicles necessary to embark on transgalactic and transcosmic (and transdimensional?) travel. This latter type of resolution of conflict is suggested in United Nations treaties signed by representatives of individual nations, and U.N. General Assembly statements. (See *Cosmic Commons* for in-depth discussion.) U.N. documents present an international consensus on the topics discussed. They are, at the very least, ideals. Some of them, in whole or in part, are legally binding.

The starkly different cultures traversing the cosmos might, much as in the classic and succeeding *Star Trek* stories, have a periodic clash of civilizations, as confederations of the malevolent encounter confederations of the benevolent. In fact, since the aggressive malevolent would find it difficult even to work together, such species might be "loners" in space, rather than be organized with other, like-species.

We do not know, given our relatively primitive technology and stage of social development as a species, which of the scenarios presented is most likely to be the reality. A "precautionary principle" put into practice at Contact in this regard would be to hope for the best (pondering possibilities to set the stage for a positive close encounter at Contact) but prepare for the worst, to the extent possible (a civilization two billion or more years more evolved than humankind would be far superior in military firepower).

In our TI reflections on possible modes of TI-ETI interaction, we would do well to realize that ETI likely shares many of our concerns and fears when they are at the same or an earlier level of technology and industry. Just as both xenophilia and xenophobia permeate human thinking about impacts of Contact so, too, is it likely present in ETI consciousness. Our attitude toward arriving aliens, then, should not be an unreserved, unqualified xenophilia *or* xenophobia. We should be open to the possibility of planetary and cosmic association and cooperation, but be prepared to try to deal with conflict.

"God's Will Be Done in the Heavens as on Earth"?

The Lord's Prayer, or "Our Father" in the Christian Scriptures (Christians' "New Testament"), praises God in *heaven* and prays that God's will be done "on Earth as it is in *the heavens*." The Latin version of the text is helpful to illustrate this: "*Pater noster qui es in caelis, sanctificetur nomen tuum* [Our Father who is *in heaven*, holy be your name]. . . . *Fiat voluntas tua, sicut in caelo et in terra* [Your will be done on Earth as it is in *the heavens*]." In the prayer, *caelis* means "heaven," and *caelo* means "heavens." A similar distinction between the singular and the plural is made in the Greek version of the prayer. (Greek was the Christian Scriptures' original language, with the possible exception of Matthew's Gospel.)

As in other cultures *caelis*, "heaven," probably refers to a specific divine dwelling place high above the world, while *caelo*, "the heavens," referred to an intermediate area above Earth and below the exclusive divine region, possibly also a divine abode (sometimes, solely for lesser deities). Principally, however, it is the locus of divine intent and action, and the place in which different types of weather develop and emerge. This distinction of realms of being is expressed biblically well before the time of Jesus. Yahweh is in heaven, above the heavens that are separate from the divine abode and in which Yahweh's creative activity is done. Genesis states that "in the beginning, God [who is in heaven] created the heavens and the earth" (1:1).

Similarly, in the manna story in Exodus, the original Hebrew text says that God gave the Israelites, who were wandering in the wilderness, manna (often equated with bread) from "the heavens" (16:4). In the account in John's Gospel, the original Greek states that God "gave them bread from the heavens"—but this became "bread from heaven" in English, a not so subtle theological shift on the part of English translators. In the original, the manna came from the heavens as a catalyzed natural occurrence, just as weather does; in the poor English translation, the manna comes from heaven, the divine dwelling place above the heavens, as a special divine creative act.

You may be wondering now, "What in the world—or out of the world—does this have to do with ETI?"

It relates back to the European Christian Doctrine of Discovery, by which European explorers and colonizers rationalized their appropriation of Indian peoples' territory and natural goods. As discussed earlier, Discovery has become enshrined in Euroamerican law and judicial practice (no longer with overt religious claims) and adopted internationally. An interstellar or at least intergalactic problem of justice would result if humans exploring outer space were to carry on this unjust tradition, and use a Cosmic Discovery Doctrine to similarly seize—steal—territory and goods of less technologically developed and less politically and militarily powerful ETI. Should that happen, the alien human invaders would oppress indigenous populations, and devastate other worlds' environments. Humans would do in the heavens what they have done on Earth. They would do to others what they have feared others would do to them, instead of doing for others what they would like others to do for them.

If humans enact Discovery in space, native inhabitants on other worlds will have their worst xenophobic nightmares come true when alien human invaders arrive (as happens in *Avatar*). Humans, for their part, will have neither xenophobia—they are technologically superior, and need not fear natives—nor xenophilia—they will be devastating ETI and ETI's home. One might hope that people with at least a minimal understanding of the projected benevolent characteristics of a divine being will notice that they are not doing God's work "in the heavens as on Earth," and consequently reject Discovery Doctrine's imperialist practices. Humans who are aware that xenophobia and xenophilia are in the minds of both TI and ETI, in *terra* and in *caelo*, will not become what they fear ETI will be on arrival on Earth, but will be what they hope ETI will be when they arrive.

Hidden Hand or Helping Hand? Equitable Economics

In the eighteenth century Adam Smith, writing in Scotland and England, provided the bases for the development of capitalism. In the same century Thomas Paine, a fiery writer born in England who became an orator and pamphleteer of the thirteen American colonies' revolt against England, provided from New England numerous, probably unintentional, counterpoints to Smith's theories in Old England. Smith and Paine, although sharing in common birth in Great Britain, envisioned far different societies and distinct foundational social structures—and the irreconcilable ideologies that they would embody. The social consequences of their respective ideas have impacted in contrasting ways people in the wealthy and impoverished socioeconomic classes in the centuries since they were formulated. (Generally in their historical era, "rich" and "poor" meant the nobility *vs.* the common people, and the owners of factories and shops *vs.* the workers in industrial and commercial enterprises.) Smith claimed that an "invisible hand" unwittingly used by the well-to-do would help the poor live on almost the same level as the wealthy; Paine declared that a helping hand from government was needed for the common person to rise from poverty and provide for their wellbeing, and to control the avarice of the wealthy. (The division between the superwealthy and the middle and lower classes was not in the same proportion as today's "1 percent vs. 99 percent" in the U.S., since the very rich in Great Britain comprised less than 1 percent of the population. However, the sentiments, ideas, and core policies for economic justice expressed by Paine and the Occupy Movement are the same: share the wealth.)

In 1776, the year in which Smith published *Wealth of Nations*, the Declaration of Independence proclaimed a vision and program that contradicted and conflicted strongly with Smith's. (Quotations from Smith's work are from The Modern Library's edition of *Wealth of Nations* [2002].) Smith envisioned a stratified society in which the wealthy maintained their position and, even without thinking about or wanting it, provided support for the common folk via an "invisible hand" that made things more palatable or at least endurable by the poorer segments of society. The Declaration envisioned an egalitarian society in which individuals would provide for themselves under a type of governance they developed themselves. Ordinary folks would need no "invisible hand" to make things more equitable, since their proposed government laws would ensure that the wealthy did not secure for themselves ever greater wealth, while the poor awaited handouts to be provided by an invisible hand—but never were. The poor did not want tenuous and capricious handouts from a hand they did not see, but visible wages received in their own hands that enabled them to provide for themselves without waiting expectantly for the accidental and begrudging sustenance that might or might not come from wealthy entrepreneurs. Smith for his part claimed that despite social and economic inequalities that would be produced by his economic construct, the poor would not lack necessities of life. Smith proposed, in effect and without naming it, the original "trickle down economics"—the excess wealth accumulated by the rich will come into the hands of the poor by an "invisible hand," and ensure their socioeconomic survival. Redistribution of wealth from the affluent few (whose riches resulted from laborers' hard work, poorly compensated) to the impoverished many was not part of Smith's system. The rich resist parting with any of their excess wealth, Smith stated; rather, they want to *increase* it by *decreasing* availability of needed goods for the middle and lower classes. (A twenty-first-century variation on Smith's theory and its "trickle down" economics is the unsupported theory—a theory contradicted by data, in fact—that taxes for the wealthy should be cut because when the rich become richer more jobs are created for the unemployed.)

In his own words, Smith describes how an "invisible hand," mobilized unknowingly by private entrepreneurs, will effect public good by promoting, unintentionally, social wellbeing: "every individual, therefore, endeavours as much as he can . . . to employ his capital in the support of domestic industry . . . by directing that industry in such a manner as its produce may be of the greatest value, he intends only his own gain, and he is in this, as in many other cases, led by an *invisible hand* to promote an end which was no part of his intention" (emphasis added). Smith stresses individual labor and accomplishment in his writings, which is understandable given his social position as an educated professor and not one of the laborers, who were generally poor and illiterate, in his native Scotland and the rest of Europe. He states that the "invisible hand" *frequently* becomes operative; it is not always at work. Some modern-day economists and politicians have used Smith's phrase to argue against any government assistance to people in need, any government efforts to establish a living wage for workers or, almost four hundred years later in the twenty-first century, any means to provide universal health care for all citizens.

Some have tried to argue that the "invisible hand" is some sort of divine guidance. They ignore that in Christian traditions God is solicitous of the poor and that the Bible prompts its readers (in the Last Judgment story in Matt 25:31–46, for example) to aid the poor directly. When the marketplace is controlled by those with disposable income, it cannot be influenced by the poor who are expending their subsistence income for whatever basic nourishment, clothing, and shelter are available to them. (The fact that poverty is rampant in the U.S. and other industrialized nations indicates that there has not been in the past, and does not exist in the present, an unconscious human or conscious divine "invisible hand" that offsets the excessive accumulations of the affluent.)

Smith claims, too, a divine establishment of class status, hierarchy, and disparity: "providence divided the earth among a few lordly masters." His assertion is in direct violation of biblical texts, in which there is no division of Earth among "lordly masters" but, to the contrary, Yahweh gives the land to the community as a whole, while remaining its only absolute "owner." At the same time, Yahweh requires a periodic Jubilee Year in which land reform is undertaken: land is to be redistributed among Israel's families so that its benefits are shared by the community as a whole and in its component parts.

The common people's economic theorist was Smith's contemporary, Thomas Paine. This compassionate revolutionary elaborated on the requisite role of government to provide for the needs of the poor, and advocated property ownership for all as a natural right. His perspective stands in sharp contrast to Smith's views. (Information about Paine's life and work are from Philip Foner, editor, *The Life and Major Writings of Thomas Paine* [1974].)

Thomas Paine was born and raised in England, in a working-class family, and was primarily self-educated. He came to know firsthand the plight of the poor. With his upbringing, experiences, and participation in the American Revolution as a background, Paine wrote *The Rights of Man, Part Second* in 1792. He declares that "when it shall be said in any country in the world, 'My poor are happy; neither ignorance nor distress is to be found among them; my jails are empty of prisoners, my streets of beggars; the aged are not in want, the taxes are not oppressive; the rational world is my friend, because I am a friend of its happiness'—when these things can be said, then may that country boast of its constitution and its government."

In contrast to Adam Smith, Thomas Paine did not pronounce a providential establishment of a classed society, but stated in *The Rights of Man* that "there ought to be a limit to property" and the wealth of vast estates is "a prohibitive luxury" if it exceeds what is "necessary or sufficient for the support of a family." In *Agrarian Justice* (1796), Paine declared that "the earth, in its natural, uncultivated state, was, and ever would have continued to be, *the common property of the human race*; in that state, every man would have been born to property" (emphasis in original). In obvious disagreement with Smith, he states that property in land should not exist in perpetuity, since "man did not make the earth . . . neither did the Creator of the earth open a land office, from whence the first title-deeds should issue." Paine advocated government intervention to eliminate inequitable property arrangements, since (and here is an obvious contrast to Smith's idea of an "invisible hand") if left with a choice regarding whether or not to provide for the needs of the poor, the rich would be unwilling to act justly: "with respect to justice, it ought not to be left to the choice of detached individuals whether they will do justice or not." In the meantime, "The great mass of the poor in all countries are becoming an hereditary race, and it is next to impossible for them to get out of that state of themselves." The rich increase their wealth, but "all accumulation" of "personal property" that does not result from "what a man's own hands produce" is derived from living in society, and any who accumulate personal property owe "a part of that accumulation back again to society from whence the whole came. . . . The accumulation of personal property is, in many instances, the effect of paying too little for the labor that produced it; the consequence of which is that the working hand perishes in old age, and the employer abounds in affluence." Since the wealthy will not give back voluntarily to society part of their "accumulation," government must ensure that socioeconomic equality will characterize a democratic society.

In the centuries following the publication of Smith's book, his theories were accepted by the dominant minority that controls economic and political structures within individual nations and on a global scale. Paine's vision and his ideas about poverty and property and concern for the vast majority were largely ignored. Capitalist ideology and economics based on Smith's theories promote the affluence of the rich and the dismal plight of the poor. Paine's alternative economic theory and structures that embody it, and government advocacy of the needs and rights of the common people, are lacking even in the "democracy" that illustrates some of his theories.

Smith, Paine, and ETI

The contradictory perspectives of Smith and Paine surface not only in humankind's considerations of what type of economics and what sort of government policies and regulations should characterize the U.S. and other nations today, but in what might result if ETI studies the extent of social concern and commitment of humans, each toward the other, as an indication of humankind's "love of neighbor." ETI assessment of economic structures and their impacts would be particularly relevant if ETI had evolved culturally to be community oriented toward the vast majority, rather than solicitous of an affluent minority. In fact, they might have culturally evolved to be an egalitarian society with an economic philosophy that presents as a social goal "the greatest good for all" rather than "the greatest good for the rich" (Smith) or "the greatest good for the greatest number" (Jeremy Bentham). ETI might use economic egalitarianism as a criterion for measuring the social maturity of an intelligent species, and thereby an indicator of how an intelligent species might "fit"—or not—in a cosmic community. If individualism, greed, ruthless competition, and lack of compassion for the "least ones"—all of which are characteristics of capitalism—prevail on humans' home planet, then humankind would be judged as likely to do in the heavens what it has been doing on Earth. In order to be accepted into cosmic community, then, humanity would need a change of *consciousness* that prioritizes needs over wants, and a change of *conduct* that embodies the converted consciousness in egalitarian economic structures and heartfelt compassion for needy others of every ethnicity.

Should this type of ETI assessment indeed take place, the dominant, affluent minority that controls and coerces many nations—militarily, politically, and economically—to benefit primarily themselves would justifiably have a degree of xenophobia in regard to their relationship with ETI. The powerful "other," the alien, the community- and collaboration-oriented ETI that Contacts humankind on Earth might propose, as a condition of consociation, that Earth be characterized by like communal and egalitarian structures. This would be simultaneously a rejection of the outcomes of the original European-Euroamerican Discovery Doctrine and of Adam Smith's ideology, a presentation of ETI's own Discovery Doctrine, and an affirmation of the communal orientation of native peoples which was rejected and substantially eradicated by European colonizers and their U.S. successors.

When ETI evaluates whether or not humankind is mature enough to be Contacted and invited to be part of a cosmic commonweal, a principle criterion is likely to be the extent to which humans have come to treat each other and their entire species in an equitable manner. ETI would recognize, as do most members of humankind who constitute a powerless majority around the world, that a sense and practice of community, which benefits all people and therefore all individuals, rather than an ideology of individualism, which benefits a few individuals at the expense of the many, would stimulate and sustain human wellbeing and be the basis for a positive assessment of humankind by other intelligent species.

U.N. Documents: Cosmic Rebirth of Thomas Paine's Ideals and Vision

Thomas Paine's statement and understanding that the Creator didn't open a land office on Earth to sell people's common heritage as individual property parcels can be extended to cover human conduct in space. This would definitely and definitively stand in opposition to transnational Earth corporations' efforts to become transgalactic cosmos corporations, and their attempts to have corporate land grabs in a subversive exercise of Discovery in space. News stories describe already corporate efforts to develop private spacecraft, in particular robotically controlled vehicles equipped with industrial technology, to claim and mine distant worlds. A recent stimulus to such efforts, already underway, has been NASA's discovery of a planet that seems to be almost entirely a massive diamond.

Several United Nations documents promulgated in the twentieth and twenty-first centuries propose that human conduct in space in the future should be guided by egalitarian principles that promote the wellbeing of all Earth's nations and peoples. Unconsciously, they recall and reflect ideas advocated by Thomas Paine in the eighteenth century.

The general consensus at the United Nations and among its member States (at least publicly) is that amid all the strife afflicting nations and peoples, ecological wellbeing and human rights are areas in which substantial agreement can be reached. The four documents summarized below—two in regard to human rights and conduct on Earth, and two to guide human conduct in space—envision and present ideals for Earth and cosmos that would mutually benefit TI and ETI, as noted.

U.N. Terrestrial-Oriented Principles and Rights

The Universal Declaration of Human Rights (1948) is an international attempt, in the aftermath of World War II and the Holocaust, to promote peace among nations and peoples, and to establish principles to which there would be global assent and commitment, even if they are not immediately implemented. The principles reflect historical contexts and realities, and affirm common aspirations for a better global future for all peoples. Its Articles present human rights thinking and principles about which the signatories are in agreement.

The rights proclaimed by this U.N. instrument include the following: "all human beings are born free and equal in dignity and rights"; all people are entitled to all the Declaration's rights and freedoms, regardless of distinctions of "race, color, sex, language, religion, political or other opinion, national or social origin, property, birth or other status," or based on any geopolitical differences; slavery and torture are banned; all citizens have equal status in legal contexts, and freedom both within national borders and external to them to seek asylum; all have a right of property ownership, individually or "in association with others," rights to freedom of "thought, conscience, and religion," and the right to participate in government, "directly or through freely chosen representatives," whose authority depends on the "will of the people"; rights to social security, to work, to choose their job and have good working conditions and "protection against unemployment," to "equal pay for equal work," to "just and favourable remuneration," to "form and join trade unions for protection of interests," to an adequate "standard of living" to provide self and family with essentials needed for "health and well-being"—which include food, clothing, housing, medical care, necessary social services, and the right to security in the event of unemployment, sickness, disability, widowhood, old age or other "lack of livelihood in circumstances" over which they have no control. The rights cited have an associated responsibility: each person must fulfill "duties to the community."

The U.N. Declaration on the Rights of Indigenous Peoples (2007) supports the efforts of the most ignored ethnic and cultural group—indigenous peoples—to secure human rights in territories and on lands where they have lived for millennia, and to seek redress for historical violations of their human rights. (Think about what benefits the rights that follow would provide in TI-ETI relationships, on Earth and in the heavens, whether TI or ETI is the species on whose world the other species arrives as aliens.) The

Declaration states that indigenous peoples "have the right of self-determination" and are empowered to "freely determine their political status and freely pursue their economic, social, and cultural development." They have a "collective right to live in freedom, peace and security as distinct peoples," and States should establish "effective mechanisms" in order to provide redress for any past action that had "the aim or effect of dispossessing them of their lands, territories, and resources," and to prevent such action in the future. Indigenous peoples, too, have the right to "maintain and develop their political, economic and social systems or institutions," and to securely exercise traditional subsistence and development practices and economic activities; the right to ongoing spiritual relationships with their traditional areas and in regard to the natural goods ("resources") existing there, intergenerationally; the "right to the lands, territories and resources which they have traditionally owned, occupied, or otherwise used or acquired," a right all States should recognize—such recognition should include respect for indigenous peoples' "customs, traditions and land tenure systems." They have the right to redress violations of their rights to land and natural goods, and redress can include restitution or at least a "just, fair and equitable compensation" for traditional lands and natural goods that were "confiscated, taken, occupied, used or damaged without their free, prior and informed consent"; if indigenous peoples' consent, the compensation may be monetary or made with "lands, territories and resources equal in quality, size and legal status." A complement to this is the right to decide and put into place their own priorities for use of their land and natural goods. Native peoples have, too, the right to receive from States and their successors recognition and enforcement of, and honor and respect for, treaties and other OSTs and arrangements that have been made. The U.N. and its agencies shall assist in the realization of the vision and goals of the Declaration on the Rights of Indigenous Peoples, including by "financial cooperation and technical assistance," and by establishing means to ensure that indigenous peoples participate in resolution of issues.

U.N. Extraterrestrial-Oriented Visions and Principles

Discussion of U.N. documents focused on human activities in space will be limited to select significant documents: the Treaty on Principles Governing the Activities of States in the Exploration and Use of Outer Space, Including the Moon and Other Celestial Bodies (commonly known as the

Outer Space Treaty); and the Declaration on International Cooperation in the Exploration and Use of Outer Space for the Benefit and in the Interest of All States, Taking into Particular Account the Needs of Developing Countries. (Dates for and details from U.N. exoEarth-related documents come from *United Nations Treaties and Principles on Outer Space*, published by the United Nations [2002].) In these and subsequent U.N. space documents, provisions that apply to the Moon apply additionally to all "celestial bodies." (Readers could consider for themselves, while reading through the U.N. texts, what might be their TI-ETI implications and applications.)

The expectations of the signatory States, according to these documents and expressed in them to varying degrees, were that extraterrestrial exploration and use would enhance human wellbeing, promote international cooperation, and help develop among nations mutual understanding, good relations, material benefits (particularly to economically poorer countries), and peace.

Proposals, policies, and principles elaborated in these Earth-originated statements are relevant for both Earth and exoEarth places. In the future, themes, ideas, and principles that originate on and from extraterrestrial places might emerge, whether through the creative, contextual thinking of Earth explorers and colonists who have traveled afar, or from the philosophies, perceptions, and practices that other intelligent beings have developed in their own contextual experiences: on their home planet, in space, or on celestial bodies.

The Outer Space Treaty was the first major international instrument intended to guide member States toward collaborative and congenial rather than competitive and combative exploration and colonization of space. Underlying this effort was global concern that the U.S. and U.S.S.R. would use their technological advantage over other nations to secure for themselves benefits in space—such as territory and natural goods—that United Nations members considered a common human heritage and province, and that places in space might become a battleground for the two superpowers. In that case Earth, too, would become a place of extraordinary conflict, a planet imperiled and assaulted by weapons wielded in and from space—including from orbiting, weapons-bearing satellites—and out to space in reply, and therefore subject to retaliatory response in turn, in unending succession until either or both sides had been reduced to ashes, abandoned cities, and barren landscapes—along with neighboring nations and peoples.

The Outer Space Treaty declares that

- the exploration and use of outer space "shall be carried out for the benefit and in the interests of all countries, irrespective of their degree of economic or scientific development, and shall be *the province of all mankind*" (emphasis added);
- outer space "shall be free for exploration and use by all States without discrimination of any kind, on a basis of equality and in accordance with international law, and there shall be free access to all areas of celestial bodies";
- no nation may claim sovereignty over any part of outer space, whether by occupation, use, or other means (note the underlying repudiation of Discovery Doctrine claims);
- all activities undertaken in space must be in accord with international laws, including the U.N. Charter, in order that peace, security, and cooperation among nations will result;
- nations shall not put nuclear weapons or other mass destruction weapons in an Earth orbit, nor shall they have military bases and installations, test weapons, or undertake military maneuvers on exoEarth places, with the exception that military personnel are permissible if they are engaged solely in scientific work;
- astronauts of all nations should be mutually hospitable, and assist anyone who is injured and away from their nation's base to return to that base;
- researchers should inform each other or the U.N. Secretary-General about any space phenomena discovered that might endanger the life or health of astronauts;
- States are responsible for damages or harm caused by objects they launch into space;
- any object sent into space belongs to the country that launched it, no matter where it is as it traverses space or where it lands, whether on a celestial body or back on Earth;
- "cooperation and mutual assistance" will guide the conduct of nations' explorers;
- "States Parties to the Treaty shall pursue studies of outer space, including the Moon and other celestial bodies, and conduct exploration of them so as to avoid their harmful contamination and also adverse

changes in the environment of the Earth resulting from the introduction of extraterrestrial matter and, where necessary, shall adopt appropriate measures for this purpose";

- nations should exercise research protocols to ensure that experiments done by one country do not negatively impact those of other countries;
- States should allow other States to observe launched objects as they traverse space;
- States should inform the U.N. Secretary-General, the general public, and the scientific community of their space activities;
- on celestial bodies, any nation should be permitted to visit all structures and vehicles of any and all other nations; and,
- the Treaty's provisions apply to all activities on celestial bodies, whether conducted by one or several countries.

The Declaration on International Cooperation in the Exploration and Use of Outer Space for the Benefit and in the Interest of All States, Taking into Particular Account the Needs of Developing Countries (1996) was issued by the U.N. General Assembly to contextualize the principle that "the exploration and use of outer space, including the Moon and other celestial bodies, *shall be carried out for the benefit and in the interest of all countries*, irrespective of their degree of economic or scientific development, and *shall be the province of all mankind*" (emphases added). It describes several considerations that should be preeminent in international space exploration. First, international law, including as expressed in the U.N. Charter and the OST, should be followed in space exploration and used to ensure that space activities are undertaken for "peaceful purposes" and to benefit all States, whatever the extent of a nation's "economic, social or scientific and technological development." All territory and the goods therein should be the "province of all mankind"; space-venturing States' special concern should be to meet the pressing needs of developing nations (we might note here that the ideas of Thomas Paine take precedence over those of Adam Smith). Second, States will freely decide how to explore and use space in an internationally cooperative, equitable, and "mutually acceptable" way; this includes ensuring that contracts for cooperative ventures are fair, reasonable, and comply with the rights and interests of all involved parties, including by safeguarding intellectual property. Third, all States should promote equitable and "mutually acceptable" international cooperation, conscious of benefiting "developing countries and countries with incipient space programmes" whose people work with nations whose space capabilities are more developed. Fourth, international cooperation must utilize the best means available—governmental or nongovernmental, commercial or noncommercial, for example—to promote participation by nations at different development stages. Fifth, cooperating countries, conscious of developing nations' needs, should have goals such as joint development and application of space science and technology, promoting space capabilities in other States, and exchanging among States, in a mutually beneficial manner, space expertise and technology. Sixth, agencies providing development aid, developing countries, and developed countries all should consider possible use of space applications to achieve their development goals. Seventh, the role of the Committee on the Peaceful Uses of Outer Space should be strengthened, including relative to exchanging information regarding international cooperation in space exploration and use. Eighth, all States should contribute to the U.N. Programme on Space Applications and to international cooperative space activities, according to the extent of their own capabilities and of their participation in such activities.

U.N. and ETI

U.N. documents published to date present socioecological (social justice integrated with environmental wellbeing) ideals that are expressed in practical principles which, if followed, would ensure peaceful and collaborative international cooperation and even TI-ETI consociation in space. They could enhance efforts to provide significant scientific knowledge and material goods benefits that would stimulate an equitable sharing of natural goods globally and cosmically, and elevate the global TI condition and if needed, the ETI condition throughout the cosmos. A simultaneous, dialogic, TI-ETI progressive development and implementation of complementary modes of thought and action would enhance the chances of achieving socioecological wellbeing in places where either or both dwell.

The U.N. documents have an anthropocentric bias in terms of human relationships with other biota and Earth, and an anthropocentric lack of consideration of the possibility that there might be other intelligent beings exploring and colonizing the cosmos. This is understandable, perhaps, in an era when humans were very self-assured in assuming and asserting that they were the only intelligent life in the universe (a claim that many people continue to make today). The documents claim that space and "celestial bodies" are a "human province" and are to be shared human property; and, they provide principles to guide space utilization in a way in which the cosmos and its natural goods could be equitably shared among all Earth's peoples. Even the documents' concern for those countries that are not as technologically developed, politically powerful, and economically prosperous as are others carries an assumption that all territories and natural goods found in space will be divided and shared by human national entities; no other intelligent being is expected to be present when humans arrive—or even afterward, since there is no mention of other intelligent species that might possibly be engaged in similar space exploration. The documents might well prove to be, despite their rejection of one nation's or one people's claims to territory and goods, an Earth-based and -biased humans' exoEarth Discovery Doctrine writ large. Humankind and human culture are preemptively presumed to be "superior" to other intelligent species and cultures, much as occurred when unequal technologies and populations clashed in the Americas and elsewhere around the globe as Europeans (and later Euroamericans) seized native peoples' territories and natural goods.

Despite their shortcomings, the documents serve as bases for pondering possibilities to enhance human beings' consciousness and conduct prior to and during space journeys and ventures. They will be helpful to develop consideration and exposition of humankind's extraterrestrial responsibilities in space exploration and settlement and, linked to terrestrial responsibilities discussed previously, a related and integrated terrestrial-extraterrestrial consciousness, commitment, and conduct.

The Outer Space Treaty and the subsequent Agreement Governing the Activities of States on the Moon and Other Celestial Bodies (1968) propose, on the whole, workable initial insights and principles for ecologically responsible human conduct—on Earth and in outer space. The rights of "humankind," for example, could be amended to the rights of "all intelligent beings"; and, statements regarding safeguarding the needs of developing countries might be rephrased to include not only their needs, but the needs of similar extraterrestrial beings, and additionally of evolving biota and evolving intelligent life. In these considerations, the collaborative development of an exoEarth consciousness and conduct could draw together the complementary perspectives of religion and science, much as ecology has done in Earth settings.

The U.N.'s efforts could have positive impacts on and implications for Contact, should that occur, and for a more "universal," rather than merely global, just distribution of control over newly found territories and natural goods, and interaction with resident biota—intelligent, evolving toward intelligence, or seemingly incapable of having or developing intelligence. Current human consciousness and conduct on Earth, and some human attitudes apparent in U.N. documents, would provoke conflict with extraterrestrial intelligent species, or at least erect barriers to congenial Contact and collaboration on Earth and elsewhere in the cosmos. By contrast, many of the U.N.-developed ideals, principles, and policies could prompt alternative thinking, and proposals for acting that would promote mutual understanding and efforts to interact respectfully in space, if those in Contact are open to responding to each other's needs.

Ideas based on U.N. statements would provide a beneficial theoretical foundation for an eventual and constantly evolving *Cosmic Charter* that would incorporate elements of Earth- and space-related U.N. documents, and thereby supplement and complement these international predecessors and the precedents they provide. Such a future Cosmic Charter could amend and extend relevant sections and ideas from earlier documents. "All States," for example, could become "All Intelligent Beings' Communities," and "Developing Countries" could be changed to "Developing Cultures." (A proposal for a draft of a Cosmic Charter is presented in *Cosmic Commons*.)

Military, industrial, and commercial objectives that stimulated initial U.S. and U.S.S.R. space ventures continue to prompt space exploration today, although in an apparently collaborative rather than conflictive way. Over the years, other nations have been incorporated into space efforts and have become involved in joint projects, such as the International Space Station, by providing scientific personnel, shuttle vehicles for cargo and personnel, construction materials, and scientific instrumentation or components needed for the operation of the station and for scientific research and experimentation. Civilian contract employees and space vehicles have been employed in space projects if only, at this point, to shuttle cargo and personnel. This might save nations' financial resources for other purposes, but it might also lead to abuse of private corporations' power, and corruption of government personnel.

Current ventures embodying cooperation for mutual benefit, and commitment to a better common future, are hopeful signs for development of more extensive collaboration projects. Their integrated operation represents a slight but significant change in human consciousness and conduct. They might not only herald humans' theoretical and practical cosmic sensibilities, but also, should Contact with extraterrestrials occur, highlight similar existing or potential concepts and constructs that would catalyze cooperation. (There have been unofficial reports that state—and at least one 2013 short YouTube video that purports to show—unidentified flying objects passing within view of the International Space Station.)

It should be noted that concepts of *xenophilia* and *xenophobia* have distinct connotations for terrestrial and extraterrestrial intelligent beings. TI hopes for Contact in which, for them, xenophilia would be strengthened: the "xeno" (alien) ETI is a benevolent and compassionate species that is congenially solicitous of human wellbeing; TI is cautious and apprehensive that at Contact their xenophobia will prove to be justified: ETI is a malevolent and conquering species that seeks to subjugate all intelligent life to its service. On the "other side," ETI hopes that its initial attitude of xenophilia will be justified and strengthened by TI consciousness and conduct, but perhaps its prior experience with less advanced cultures (or its own history) has led it to a cautious xenophobia. Similarly on other worlds, where voyaging humans are ETI. On planet X, human ETI as the arriving alien, and resident TI dwelling peacefully in their home, initially would exercise a corresponding caution as each wonders whether they will share mutual xenophilia, mutual xenophobia, or have a confrontation of contrary attitudes—one with xenophilia, and the other with xenophobia

Welcoming the Stranger-Foreigner-Alien

The Genesis story of Cain and Abel (discussed in chapter 1) illustrates the ancient attitude of suspicion about an approaching stranger that is still operative today for cultures and individuals around the world. The attitude often precipitated an immediate hostile defensive action: killing the stranger. People feared that if someone were an outcast, or a possible scout for an invading army, they must be killed to prevent the stranger from harming them directly or reporting back to the rest of their invading force such information as the population, apparent number of defenders, fortifications, etc.

Archbishop Demetrios, primate of the Greek Orthodox Church in America, described the history and exercise of Christian love of the stranger-foreigner-alien in "Love and Care for Strangers: *Philoxenia* in Christian Tradition." He observes how radical was the early Christians' doctrine and practice of *philoxenia*—which joins the roots *xeno-*, "strange," and *philo-*, "love." Thus, "*Philoxenia* is literally 'love of the stranger,' 'love of the foreigner or alien.'" He quotes an 1859 campaign speech by presidential candidate Abraham Lincoln in which the contrasting, millennia-old approach is described: "From the very first appearance of man upon the earth, down to very recent times, the words *stranger* and *enemy* were quite, or almost, synonymous." Lincoln states further that even in an era of "civilized nations," it was still acceptable in some countries to harm others in ways legally prohibited among their own people: to "rob, murder and enslave strangers." Archbishop Demetrios describes the distinctive, two-millennia-old Christian tradition in which social norms of hospitality toward others are a base but not the sole factors in relating to the stranger. Christians went far beyond that, says the archbishop:

> But to truly love the stranger and foreigner, to love the one who is different from oneself—this requires a motive more pure and more profound. What we find in the New Testament is an ethos of *philoxenia* based not on fear but on care and gratitude. . . . Hospitality to the stranger is celebrated as an occasion for offering love. . . . The love of the disciple, the hospitality, extends to those who cannot reciprocate, to those even who will not reciprocate— to the poor, the stranger, the enemy and persecutor. . . .
>
> Of the many counter-cultural aspects of early Christianity, its spirit of love for strangers, its hospitality in action, would have been among the most controversial. The Christian willingness to bring foreigners into their homes, to give them aid and shelter, must have seemed to some like an act of recklessness, or utter naiveté.

In the twenty-first century some vestiges of Christian *xenophilia* remain. Individuals and religious groups welcome and care for strangers, including homeless people on streets, immigrants from other countries and ethnic groups, exiles, and escaped victims of torture and political and economic oppression. The United Nations too, in its refugee efforts, attempts to care for victims of civil wars, foreign invasions, and domestic oppression; it provides refugee camps, food, shelter, medical aid, and other necessities, sometimes to large groups of people living in camps.

So, we have two basic, contradictory approaches to the *xeno* among us or approaching us: love or hatred. There are diverse ways to express either attitude in the presence of the alien other, each of which would await some indication of a similar kind of response: a tentative and cautious initial acceptance, or a "show of force" display, at the ready to accelerate defensive action if needed. ETI presents a unique *xeno* case for our consideration: they are technologically far superior to us, judging from credible witness reports (including those of military personnel who have observed the speed and maneuverability of ETI craft from radar tracking, from visual ground observation of the skies above, and from cockpit views of military pilots who have tried to approach them and have, on occasion, fired upon them—with no hostile response, solely a rapid relocation, generally straight up at extraordinary speed), so humans' immediate hostile reception of ETI would probably not be a good idea. They might actually have evolved beyond aggression among themselves, and are hopeful that we have begun to reach a similar evolutionary stage. In that case, humans' initial neutral or congenial reception of ETI would reap a reciprocal positive response.

In light of present and past responses to ETI, a disconcerting possibility emerges about a significant ETI response in the near future. To date, the ETI reaction to human military aggression has been withdrawal. Continuation of violent TI attacks against ETI might cause benevolent ETI to leave Earth's region and not return. This ETI decision and action would open the skies to the type of ETI whom we fear. Malevolent and seemingly emotionless ETI have been reported by some who claim to be abductees, who have said that they underwent surgical procedures, and by Colonel Corso, who had been stationed in the Pentagon and who stated that all ETI are malevolent, citing as evidence human abductees subjected to medical experimentation, and cattle mutilations. We cannot afford to continue indiscriminate and irrational hostile attacks against all UFOs that come in range of human weapons.

Humans' lack of prior extended experience with Contact makes matters more difficult for us. Is ETI friendly or hostile? Are there different species of ETI, evolved on different worlds, such that some might be friendly and some hostile? Should we be initially friendly with all ETI, hoping that ETI is similarly disposed and will respond in kind, or display an obvious readiness to attack in case they are oriented toward aggression? (We should probably be ready with a fallback defensive position in any case.)

TI-ETI Contact: Xenophobia or Xenophilia?

It's great to talk theoretically and speculatively about what Contact would be like with ETI, but what might we experience at a Contact moment? Would we be fearful or joyful? Apprehensive or curious? These dispositions might determine whether or not we act aggressively or judiciously when TI-ETI Contact occurs.

An aid to our considerations and conduct would be that rather than regarding ourselves as "terrestrial" (on Earth) or "extraterrestrial" (on exoEarth worlds) in any or all contexts, we will recognize that all of us have a common cosmogenesis in a cosmic community of communities, each of which has its origin in distinct cosmos contexts. Intelligent beings are gradually coming to be aware of, and might eventually know interactively, other members of a (potentially) common community that is related through the reflective consciousness of each evolved form of intelligence.

We are a community of conscious stardust. We are thinking atoms. We are part of the complex cosmos, conscious of itself holistically through the integration, interdependence, and interrelation of distinct parts, most fully in and through all intelligent beings.

The extraordinary event of interpersonal Contact between terrestrial and extraterrestrial intelligent life in a dramatic public way, if it occurs and is acknowledged, would have substantial impacts on the distinct species involved. It might well alter the ways in which we humans understand or envision ourselves not only in our particular Earth home in our specific solar system, but in the immense cosmos.

Prior to such Contact, it would be well for humankind to consider their current and projected interbiotic, interplanetary, intergenerational and, perhaps, interdimensional responsibility. How should humans relate to other members of the biotic community, whether intelligent or less complex, whom they have not encountered previously? What might biotic rights be when considered in a cosmic context? How would "natural rights" be explained and elaborated interculturally and intercontextually? In such considerations and speculation, it would be helpful to develop proposals for human conduct, expressed in initially formulated principles, which would evolve in context if there were a constructive consultation with ETI on Contact, or as soon as possible after Contact. Efforts to consociate, probably tentative at first as each species seeks to understand the other and weigh concepts and principles that might not have been previously considered, should prove to be beneficial to all parties. As intelligent beings, all would be conscious of their interrelationship within integral being, however understood and expressed.

U.N. documents and, possibly, a Cosmic Charter in progress could provide initial principles for interactive conduct at Contact. In preparation for space travel and for ETI encounters in space or on Earth such a Cosmic Charter could promote changes in human consciousness and conduct—for the mutual benefit of people and ETI. The Cosmic Charter, as originally developed and elaborated on Earth, would advocate for interbiotic, interplanetary, and intergenerational responsibility. It would incorporate the best of current international thinking on three key ethical issues: social justice and wellbeing, Earth and ExoEarth ecological wellbeing, and terrestrial-extraterrestrial mutual respect and accommodation.

Reaching for the Stars

In order to promote planetary and interplanetary wellbeing, people need to develop a new sense of their place in and relations with their Earth home and in the expansive and still expanding cosmos. In so doing, humans would learn to respect both who they are and where they are going, and become responsible explorers and inhabitants throughout space and through time. People would be educated and stimulated to be socially and ecologically considerate, to be respectful during terrestrial-extraterrestrial engagements, and to develop—before and during explorations—religious or spiritual reconciliation with the implications and impacts of Contact.

Humans as a whole do not know with certainty whether or not aliens have come among them already—the Roswell Incident in New Mexico and Hudson River Valley sightings in New York, among others, are still being debated. People do know that they live in an extraordinarily dynamic and beautiful creation—the wonder of the Hubble telescope's images is not debated, nor is the intricacy and diversity of life on Earth, some of which is viewed only through microscopes.

The cosmos is a vast and complex expanse with multiple dimensions in time and space. All comprise a single reality, and are related to each other in the integral being of all that exists. The existents, energies, elements, entities, and entropy of the universe are intertwined, integrated, interrelated, and interdependent. In incompletely known and complex ways, they influence and impact each other in time and through distinct dimensions of reality.

Intelligent beings are conscious, reflective, creative, curious, and contextually adaptable inhabitants of the cosmos, in whole or in part. They have developed modes of cosmic travel that enable them to voyage expeditiously and safely through what would be, in observable space-time, immense spatial distances and distinct zones of time: past, present, and future. In their voyages, they come into Contact with other intelligent beings who comprise diverse species with distinct technological capabilities and possibilities; disparate environmental contexts on sometimes distant planets or celestial bodies; and new (to them) and innovative native cultures and the institutional (political and economic) and transcending (philosophical and spiritual) social structures that comprise and organize their relationships to each other as social beings in social contexts. Intelligent beings could consciously integrate with each other for mutual benefit, and to envision, develop, and realize common wellbeing. In newly discovered environmental and cultural contexts, too, they would become aware of diverse types of ecological integration: among members of intelligent species, between communities of intelligent species, between intelligent species and less complex species, and between all species and their planetary setting.

In their initial encounters and engagement intelligent beings—TI and ETI—should seek to develop such relationships and structures as will enhance their consequent interaction. In order to proceed pacifically and to provide mutual benefits, intelligent species should strive to effect equitable allocation of mutually sought territory and natural goods—as provided in various contexts by the dynamic cosmos—each or all of which might be necessary or useful for particular beings' organized community entities during some or many of their stages of social development: to meet the shared needs of their own societies or of other societies with which they come into Contact.

TI and ETI, if they overcome *xenophobia* because they have come to know each other's consciousness and conduct, would seek to develop an ever more inclusive community among intelligent species in the extensive and diverse places of the cosmos. Their essential *xenophilia*-become-*biophilia* will enable them to voyage in compatible and complementary ways throughout the cosmos.

Congenial, collaborative, and creative efforts of diverse intelligent beings would facilitate establishing *xenophilia* as a mutually shared attitude and bond. Eventually, through mutual recognition of the diversity, characteristics, and needs of each member and species of a cosmic community, *xeno* will become increasingly less a distinction and separation, and *xenophilia* will evolve into *biophilia*, a love for all life, expressed in ways appropriate to different levels of biotic complexity. In cosmic community, "to each according to their needs" would provide a good foundation for ongoing collaboration. Once this stage of common existence is attained, then the guiding principle would be "the greatest good for all."

In ways far more complex and extensive than he could ever have imagined, Charles Darwin's words about the development of life on Earth will become ever more true and profound for amazed voyagers into outer space. Aware ever more extensively and in depth of cosmic dynamics and complexification since the moment of cosmos birth Georges Lemaître called the "day without yesterday," they will go further back into time and farther out in space. When they ponder the origins and understand the expanse of the cosmos, they will be increasingly awed and even overwhelmed to an even greater extent than Darwin was when he understood the origins and evolution of Earth's biota. They will appreciate the universe in a way distinct from but complementary to Darwin when he was inspired to comment, "from so simple a beginning, endless forms most beautiful and most wonderful have been, and are being, evolved."

5

ETI ENCOUNTERS: STATE, ACADEMIA, CHURCH, AND SCIENCE

Adversarial and Accommodating Attitudes toward UFO Reports

AROUND THE GLOBE AND throughout history, people have recorded and reported near and distant encounters with unidentified flying objects. Credible witnesses throughout the U.S. have described unusual, unidentified aerial phenomena (UAP) that they saw. Sightings have included unknown aerial craft traveling at extreme speeds that, without stopping or banking in flight, changed direction suddenly. The changes of direction included a UFO's or UFOs' horizontal flight across the skies that went abruptly in a vertical direction, perpendicular to their original trajectory. At other times, one or several spacecraft made a series of "Z" maneuvers at high speed in the night sky. This maneuver is regarded as "impossible" by Earth scientists, who judge it on the basis of humans' current knowledge of physics and of humans' projected technology developments and energy sources. People who witnessed such occurrences remain in awe decades afterward. Although very curious about the apparent space vehicles performing these acrobatics, they remain silent about or only selectively relate their experiences, even to members of their own family, people in their social circle, friends in general, and trusted professional colleagues who are friends. They fear that the reactions to their reports of what they've seen will include personal ridicule from friends and family, and adverse professional consequences in their places of employment.

Government, Academia, and Religion Consider ETI

Officially, according to the U.S. government and military, UFOs and exoEarth aliens do not exist, and stories about them—and even photos or videos—should not be considered seriously by curious and thoughtful U.S. residents. The fantastic, unscientific government message, without any definitive or even tentative evidence: "We are alone" in the cosmos. This sentiment has been embraced by academic institutions. Religious institutions have, for the most part, fallen in line, although they still might consider appropriate responses to the appearance of and Contact with UFOs and ETI.

U.S. Government Denial: There Are No UFOs

Despite reports and data presented by courageous scientists about anomalous atmospheric and outer space phenomena that dramatically violate known laws of physics and humans' current technological instruments' capabilities the U.S. government, unlike most governments in other countries, consistently has denied the existence of UFOs and ETI. Official agency representatives often have circled around central issues such as, most obviously, evidence available in credible witness reports—including when these are accompanied by photographs, and linked to scientists' on-the-ground site analysis—that contrast sharply with the "official story," for which no valid evidence is presented. The most notorious examples in the U.S. of this type of obfuscation, which includes outright denial but also circling around and never directly engaging the evidence—at least publicly or officially—have occurred at Roswell, New Mexico, and in the Hudson River Valley, New York.

Ranking military personnel have stated for decades after the 1947 Roswell events that alien craft had indeed crashed on the Foster Ranch. However, the official interpreter of events, the United States Air Force (why is a military component and not a civilian branch of government in charge of "investigating" such events—or is it, ultimately?), has consistently claimed that a weather balloon crashed, with or without a dummy attached. The "consistency" is present only in the general type of claim ("a

ETI Encounters: State, Academia, Church, and Science

balloon crashed"). As the press finally began to investigate and probe into Roswell and press for the truth, the USAF changed its story time and again when confronted by evidence from its own data that no weather balloons were in the area, and that the nearest ones were of a different type than what had been claimed previously to have crashed. As people continued to come forward to state that they had seen aliens who had been on board the crashed craft, the USAF claimed that this was not the case, then that there were dummies attached to the balloon; but then, when USAF reps were confronted with their own data that revealed that the placing of dummies on weather balloons occurred several years *after* Roswell, the USAF abruptly stated that UFOs did not crash at Roswell, do not even exist, case closed, and furthermore the USAF will not hold any more public meetings on Roswell or any other UFO-related subject.

Hundreds of credible witnesses in the Hudson River Valley (including pilots, police officers, engineers, and an IBM executive) have reported that on several occasions—sometimes for several days consecutively—they saw UFOs fly silently and slowly overhead, passing close to surface structures. Area newspapers covered the story. The federal response, forced on local police personnel and security officers: the lights that were fixed on massive unidentified aerial phenomena showed the overflight of ultralight planes—even though high winds were blowing. Such winds would have made it at least extremely dangerous, if not impossible, for ultralights to be flying in the area. Local residents were furious about but unable to force a change in this "official story." As one outraged observer said, "Who am I supposed to believe: my 'lying eyes' or my government?" Inquisitive scientists have suggested a course of action in response to UAP that would be a reasonable alternative to government rejection without investigation: an independent scientific inquiry into such phenomena.

Pilots, astronauts, military officers, politicians, professors, physicians, and truck drivers, among others, have reported sighting—and in some cases, interacting with—alien craft. Despite all of this, the federal government has refused to provide funding for any independent research endeavor that seeks to establish objective scientific data pro or con about UFOs.

Why have successive U.S. government officials (from both major political parties), and those charged with UFO investigations who control dissemination of data acquired to date, refused to be transparent about what they have? Why don't they support independent, ongoing, open, and objective scientific investigations of past, present, and future incidents, and data about them? Shouldn't government officials want to try to establish scientifically what they have claimed for decades without proof of their position—that all UFO and ETI reports are misperceptions by otherwise credible witnesses—and disprove thereby all allegations that the government is involved in a massive cover-up operation? Do they fear that independent scientific research will prove, contrary to government assertions, that UFOs and ETI are materially real and that witnesses' and journalists' assertions are accurate—that there is a federal government (and military) cover-up of details not only concerning UFO events, but also UFO debris recovered and efforts to "reverse engineer" it to manufacture commercial, industrial, and military products?

Government- and military-sponsored and controlled press conferences have, for decades, presented only the government's official narratives about reported UFO incidents. A public relations specialist—propagandist—for the sponsors tries on these occasions to deny that people have seen what they declare that they have seen. If government "facts" do not satisfy people present who know what they have witnessed, the PR person will use condescension and ridicule in an effort to discredit witnesses and dismiss their accounts of UFO events. The government PR spokesperson often has little or no scientific background. Yet members of the general public—scientists included—who have not seen what the witnesses describe will reject it out of hand, but will accept the PR story presented by a government nonscientist who was not present at the event. Distinguished research scientists who build upon and accept the work of their predecessors without replicating every experiment and analysis—because they have faith in the credibility of their expert antecedents—will challenge every hypothesis or theory put forward by others in their field who do not have such scientific "street cred." However, when it comes to speculation on or assertions of the possible presence and impacts of ETI on Earth, they accept without question a nonscientist public relations officer's explanations of UFO phenomena but reject a witness report submitted by a fellow scientist, who surely has more credibility than the official PR rep. Uncritical and unreflective acceptance of an "official report" and simultaneous automatic rejection of a witness report together express an unscientific position and prejudice.

Why do scientists not demand of government PR reps the same sort of evidence they require from credible witnesses? Why do they not apply scientific criteria to analysis of PR statements? Perhaps their stance reveals

ETI Encounters: State, Academia, Church, and Science

an unconscious, innate fear about a more technologically advanced species, a fear that erects a psychological barrier against credible witness testimony; a conscious fear about professional punishment if they appear even to ponder UFO witness reports, let alone advocate for scientific study of phenomena; or a personal attitude that demands—including for the reasons just cited—material "proof" of witnesses' statements. It has become acceptable for people of all walks of life to accept government and military "public mind control"; to put blind faith in "official sources"; and to reject professionally and societally respectable people's reports—even when the latter have photos ("they were obviously Photoshopped") or videos ("obviously contrived"). Sometimes, because of other government cover-ups, people who claim to want evidence and "objective" testimony declare that UFOs were secret U.S. military craft whose existence and capabilities were classified for "defense" purposes—even when UFO maneuvers high above and when approached by air force pilots in attack mode, or when tracked by aerial or ground radar—or both, simultaneously—far surpassed any technology currently known, and violated laws of physics currently accepted. By contrast, when Winston Churchill received a report from the Royal Air Force toward the end of World War II that two RAF pilots returning from a bombing mission had been accompanied in the air for a time by unidentified craft with exceptional maneuverability, he ordered the file to be sealed for fifty years to avoid a public panic. Churchill did not deny that the incident had occurred, nor did he question the veracity of the RAF pilots involved.

It has become acceptable, somehow, that highly intelligent people—perhaps, at times, for the reasons cited above, or their own intellectual arrogance about the extent of their knowledge, or philosophical-religious apprehensiveness—refuse to consider the possibility of an ETI presence in the entire cosmos, let alone in Earth's atmosphere or on Earth's earth. They continue to reject, irrationally, an objective independent scientific investigation of UAP when a government propagandist with little or no scientific credentials advocates otherwise while claiming unabashedly, unqualifiedly, and without presenting evidence that UFOs and ETI do not exist. People present for UFO-related, government-sponsored press conferences or public presentations rarely, if ever, question the PR rep: "Where's the proof?" and, more importantly, "Where's the truth?" People here evidence the same flaw characteristic of religious fundamentalists, including creationists: they have already decided what is "true" about cosmic creation, and reject any alternatives to their own "truth," particularly in the form of evidence provided by scientists, historians, and other scholars. Oddly enough, fundamentalist atheist scientists here mirror the attitudes and practices of people whose perspective they otherwise reject. Fundamentalist beliefs are not characteristic solely of a particular type of theist; they are held, too, by atheist scientific materialists. These attitudes remind me of a line in Paul Simon's classic song "The Boxer": "A man hears what he wants to hear, and disregards the rest."

Government efforts to suppress UFO information and prevent UFO investigation have been supported in some cases by religious institutions, and in other cases by academic institutions. These are certainly complementary efforts. There might be some coordination of their endeavors, such as between government and academia, since universities fear they would lose needed research funds—government grants on which they are highly dependent for innovative and advanced scientific studies—if they were to explore, scientifically and social scientifically, UAP and ETI narratives.

Religious institutions have, however, sometimes independently explored the heavens and analyzed Earth incidents. On occasion their scientists have speculated about the possibility that ETI does exist, apparently has passed very close to Earth, and perhaps even has landed on our planet. Representatives of the Catholic Church, for example, whose Vatican Observatory has powerful telescopes available for its use and scientific research teams at its command—at Castel Gandolfo, just outside Rome; the Steward Observatory at the University of Arizona, Tucson; and the Steward Observatory-affiliated Mount Graham International Observatory (MGIO) near Safford, Arizona—sometimes have speculated about ETI existence. Church-affiliated scientists may well have observed dramatic UFO phenomena that were anomalous in terms of the ordinary and expected behavior of natural cosmic occurrences such as meteors.

Academic Abdication of Educational Responsibilities

Academic institutions generally have supported government rejection of scientists' suggestions that the possible existence of UFOs should be seriously analyzed. Even research universities, supposedly interested in discovering and studying diverse aspects of science, have threatened, implicitly or explicitly, faculty who seemed curious about exoEarth phenomena or who expressed a serious interest in UFO and ETI studies. Institutional official denial kicked in even when scientific data indicated that UAP seemed to be under intelligent extraterrestrial control. Individual faculty members, entire academic departments, and members of the central administration in numerous institutions have derided colleagues just because they have expressed their curiosity about UFOs, even if they have not stated that they have seen one. Sometimes this happens even when scientific researchers have collected data to support UFO and ETI hypotheses, or have had experiences of their own.

Ridicule has served the "debunkers" well, especially when it is coupled with the threat that scientifically curious faculty would lose their academic status, not receive promotion or tenure, and suffer social isolation if they sought the scientific truth about UFO phenomena. Theoretical physicists, astronomers, and biologists—including experts in the newly emerging field of astrobiology—are frustrated in their quest for scientific truth by this institutional and departmental policy. But most conform publicly, even as they reject privately the current secrecy about and sequestration of hard evidence that ETI is observing us—perhaps as part of a sociological study or scientific survey of intelligent life on worlds other than their own, or perhaps even as . . . tourists (and, for the most paranoid or "realistic" humans, as scouts for a possible invasion). It will take a courageous administration's openly expressed support of objective scientific study of UFO-ETI phenomena in order for faculty to be openly interested in such an effort—after they "come out of the closet." When institutional leadership and academic expertise in the field are linked publicly, with the result that evidence emerges of UFOs during the scientific quest for knowledge, a catalytic moment might well occur—for science and society. A pertinent comment on this issue was made by Paul Hellyer, former Canadian Minister of National Defence, in an August 2013 email to this writer: "My heart's desire is that we will have the openness to seek the truth whatever it may be. I'm old fashioned enough to believe that real freedom depends on it! In respect to the extraterrestrial presence and technology, it comprises a new and broader perspective of the cosmos, that it is teeming with life, with both animate and inanimate being. It is a fascinating area of research and discovery for anyone who is inclined to seek the wider horizon."

Religious Resistance to ETI Existence and Spiritual Autonomy

As human civilizations emerged, speculation about life on other worlds began to develop more extensively. Ancient Greeks philosophized about the possibility. In Genesis, the Hebrew Scriptures describe briefly an era when "giants walked the Earth," and the "sons of God" had intercourse with the "daughters of men." Historically, the Christian churches ordinarily have denied that ETI exists. The monk Giordano Bruno was burned at the stake in 1600 for suggesting such a possibility. Churches' attitudes and actions often have been caused by their fear that some theological doctrines would be called into question if the existence of UFOs and ETI were to be verified. As data has accumulated, however, what has transpired has been similar to Christians' reception of the theory of evolution advanced by Charles Darwin in the nineteenth century: most rejected it until the scientific evidence became overwhelming and they had to acknowledge that evolution has been occurring; some qualify evolution, teaching that it is operative for all biota except humankind, and others deny it still, despite objective evidence to the contrary.

In the book of Acts in the Christian Scriptures, in the story of Jesus' ascension to heaven, onlookers gaze skyward for a while after Jesus is no longer in sight. Suddenly, two men in white garments appear next to Jesus' followers and bring them down to Earth with the question, "Why do you stand looking up toward heaven?" (1:11). The disciples mentally descend back down to material Earth and return to Jerusalem. They realize it is time for them to continue the work of Jesus, rather than wait until his expected imminent return. On Pentecost, they had a communal experience of the Spirit, and formed themselves into a cohesive community of believers to carry on the work of Jesus. This Pentecost narrative embodies the message of the mysterious men at the scene of the ascension: carry on Jesus' teachings as a community of believers immersed in the realities of material life in this world; don't expect and await dramatic instructions or interventions from heaven.

In biblical times "the world" meant Earth and everything visible from Earth (sun, moon, stars). The skies and the heavens (not including "heaven," an exclusively divine region believed to be above both the dome of the sky and the heavens below the dome) between Earth and the upper dome were viewed as part of Earth's extensive "world." Since they did not understand the immensity and complexity of all they could see, or the vast expanse of the cosmos in which what they saw was situated at incomprehensibly great distances, they developed two distinct possibilities or interpretations of the meaning of the "world" to which Jesus instructed his followers to spread his message.

First, the "world" means Earth alone, including its atmosphere but not what is beyond. Just as the Ascension witnesses are to stop looking "above" but become engaged "below," subsequent believers are to be responsible Christians in and for Earth and all that exists on Earth. God will interact with other worlds as God decides, not humankind. Intelligent beings on other worlds might already have had a special embodiment of God as a member of their own species who instructed them, by words and example, in the way God expected them to live. In this scenario, every world's inhabitants have the benefit of divine presence among them in their own material form (if needed or helpful). Therefore, all (TI and ETI) are to be responsible intelligent beings in and for their own world and should not seek to convert other encountered intelligent beings, who have their own teachings and spiritual-material way, to only one, cosmic way—their own—of relating to the divine Spirit.

Second, the "world" continues to mean for Christians on Earth, despite two millennia of scientific and social scientific data provided since the first Christian century, everything visible. Whenever and wherever another part of the Earth-cosmos "world" becomes accessible, then and there Christians are to spread Jesus' message. However, a result of this attitude and endeavor might be that, just as happens now on Earth, a religious conflict among believers could occur: ETI might have a similar unalterable ideological perception that they should spread, throughout the visible "world"—the entire cosmos that they see in their sky—the message of God as it was given to them. When ETI viewed space in their pre-technological age they, too, might have come to believe that their "world" included all the heavenly bodies they saw. Effectively, to them, Earth—human's origin planet—is part of their planet world just as their origin planet is a part of Earth's world. In both perspectives, the "world" includes all that is visible to the stargazing inhabitants who live on a particular planet.

If ETI has received, similar to humankind, spiritual instruction to spread the religious teachings they received from someone they believe is a divine messenger to the whole "world," they would (with as much or as little justification as humans) seek to incorporate other intelligent beings into their religion (however constituted). In such an ETI perspective humans should be converted to follow ETI's ways. An unnecessary conflict of metaphysical beliefs would occur should TI and ETI both believe that only their species has teaching from a divine Spirit that they should spread throughout the cosmos, which must be accepted by all intelligent beings they meet.

Baptize ETI? Whom Would Jesus Save?

In recent history, periodically over the years, whether because of a new report about a UFO sighting, a new science fiction film, or a documentary or interview on television that suggests the possibility of intelligent life elsewhere in the universe, members of some Christian groups have responded with calls for a new type of missionary effort: to instruct and baptize alienkind into a particular—and, for them, the only "true"—form of Christianity, namely, their own. Such Christians operate with the assumption that human conduct on one planet among nine planets in one solar system, in one galaxy among billions of galaxies, indicates that all intelligent beings in the cosmos will act in like manner: they, too, will be a "sinful" species. These perspective-limited and doctrinally narrow believers say that Jesus came to Earth to "save" the "world," which they understand to be the entire universe. They will carry on Jesus' work on Earth and extend it into the cosmos of many worlds, and thereby will provide the means of salvation from "sin" in present, material worlds in order that TI and ETI alike might become saved beings who will have eternal happiness after death. Their blessed state will be effected in the future in the exoEarth spiritual world of "heaven" that they will inhabit. This spiritual salvation will be achieved through their belief in Jesus. Without this belief, they will suffer eternally for the consequences of their sins; they will be exiled to "hell," a nightmarish world of physical pain and spiritual suffering. Christians with this view project, then, onto intelligent aliens throughout the cosmos, an ETI version of humans' (im)moral conduct on Earth. They assume anthropomorphically that since humans have messed up their primordial relationship and ongoing historical relationships with God, so too, must extraterrestrials have done so in their own world. They assert anthropocentrically that God came among human intelligent beings on Earth exclusively, rather than also among other intelligent beings in the cosmos who also needed divine intervention in order to have a way to heaven. It is necessary, then, for ETI's own good that they be baptized. The Christian religious ritual requisite for all TI and ETI is not, they assert, intended to strengthen humans' faith and thereby buttress humans' credal comfort and religious security: "we have the *true* religion" (cf. Discovery Doctrine).

Christians on Earth might give just a little more credit to a loving God who, Genesis declares, views all creation, including all creatures—biotic and abiotic—as inherently "very good." God might have become part of ETI species in ways complementary to divine incarnation in Jesus. Or, at the very least, the divine Spirit inspired seers and prophets on other worlds to teach a way of life that corresponds to what a loving Spirit hopes for them, and guides them toward. Religions or spirituality might or might not have developed after the inspired leader passed on. Consequently, ETI does not need human baptism into a way of life that, traditionally for Christians, was provided in an Earth context for humankind. In fact, ETI might even have evolved to a higher spiritual understanding in which institutional forms of religion have become irrelevant; the original teachings that were part of their spiritual formation and foundation had endured not in competitive religious institutions but via their incorporation in some form in the perspectives and lives of ETI.

Baptize ETI? No. Rather, engage ETI in an open and interested quest to learn their form(s) of spiritual practices, if any, while conveying humans' metaphysical understandings, as expressed in institutional religions' doctrines, secular humanist thought, and other social and individual ways. Mutually instructive forms of metaphysics and spirituality might result from such TI-ETI encounters and conversations.

Cosmo-racism

Ethnically distinct members of the human race experience racism at the hands of and in the discriminatory laws promulgated by particular segments of the societies in which they live. Racist comments notwithstanding, we are all members of the same human species and should be a familial community. Africans, Asians, Italians, Irish, and Mexicans, among ethnic groups in the U.S., have been oppressed by the dominant culture developed by Europeans and Euroamericans beginning in the "Age of Discovery" when the Eurocentric Discovery Doctrine was originally expressed. Subsequently, it was elaborated in U.S. contexts by Euroamericans and imposed on Indians, the original human inhabitants—the original American natives and native Americans—of the Americas.

In light of this history, what might anatomically, intellectually, and socially distinct ETI experience on arrival in the U.S. and elsewhere around the globe, or on distant worlds? Or, for that matter, humans arriving on distant worlds? Is cosmo-racism a possibility? The movie *District 9* explores this issue to some extent as it presents human attitudes toward, and practices against, the "prawns" who arrive on Earth—policies and practices that mirrored what transpired in South Africa during the apartheid era, and in the U.S. South not only when slavery was illegal, but in Jim Crow laws, lynchings, and housing and employment discrimination for more than a century after the Civil War. During that conflict, Abraham Lincoln had issued the *Emancipation Proclamation* to free the slaves, a proclamation subsequently passed by Congress to become federal law.

Human cosmo-racism will certainly be constrained if technologically (and therefore probably militarily) superior aliens arrive on Earth from outer space, or TI aliens arrive on worlds on which culturally and technologically advanced ETI natives dwell. Even better: humankind should eliminate its racism of whatever kind, its economic oppression of the weak of its own species, and its militaristic conquest mentality before departure to other worlds. Humankind must convert its current assertive and discriminatory consciousness and conduct into compassion for intelligent beings and all biota, and concern for its planetary contexts.

The prospect and danger exist, too, that ETI will have its own version of cosmo-racism. However, we cannot speculate about ETI attitudes before evaluating and altering our own. We must, in a biblical analogy, remove the actual beam from our own eyes before trying to remove the theoretical splinter from ETI eyes.

Evolution and ETI: In the Heavens before It Was on Earth?

Scientists estimate that the universe is 13.8 billion years old (okay, that's a pretty precise "estimation"), that Earth is 4.5 billion years old, and that life emerged on Earth about 3.5 billion years ago. What about planets in the rest of the universe, increasing numbers of which are being discovered weekly, it seems, by the Hubble and Kepler telescopes? How many of these are in the "Goldilocks Zone," orbiting far enough away from their star to be neither too cold nor too hot to support life as we know it? How many planets "in the zone" have conditions hospitable to life's emergence, and on how many of these did life actually emerge? To what extent is emergent life evolving toward (or has life evolved already to become) intelligent life? As you can see, merely locating a "Goldilocks" planet is not sufficient evidence that we, as thinking beings, "are not alone." Note, too, that humans will assess these questions on the basis of their current human understandings of what "life" is and what "intelligent life" is.

It is a given among scientists and those who appreciate science that once life has evolved it complexifies and diversifies to some extent. Depending on its material context, life might or might not complexify to the extent of including intelligent life. It is possible, too, that there exist types of "life" significantly distinct from life on Earth, whose existence might not be detected at all, or until we understand that we have come upon a different type of biotic being. On Earth alone, we've already discovered organisms living in extremely hot places deep in the ocean where volcanic activity creates temperatures previously believed to be inhospitable to life, and other organisms living in extremely cold conditions in the Arctic that previously would have been theorized to be a hostile environment for biota.

A complementary speculation and consideration: does intelligent being always have biotic materiality? Might it have another form of *be*-ing? Consciousness and intelligence on other worlds or galaxies might be distinct from what they are on Earth. Conceivably, it might be more accurate to speak generically about intelligent *being*, not solely about intelligent *life*: we do not know, nor should we anthropomorphically project, what intelligent beings might be like throughout the cosmos. Where intelligent life does exist, we can assume that it evolved to its intelligent state and was neither a specific divine creation nor a sudden spontaneous emergence from existing matter. Evolution in the heavens might or might not have occurred as it did on Earth.

Might it be possible that since the universe is more than three times the age of Earth, evolved life is not only far superior to humans in intelligence and technology, but also that its evolution has brought it to a higher state of consciousness? If so, might its materiality, expressed in terms comprehensible to humankind, be minimal and its far more complex brain (or parallel thinking organ) have brought it to a much higher level of spiritual consciousness and activity, and perhaps to a greater presence within or participation in spiritual dimensions of reality?

We do not know, and few care to project, toward what similar state humans might be evolving. Consideration of TI or ETI evolution into a less material, more spiritual mode of being seems to threaten the ideologies and psychological and intellectual security of both those who do not believe in spiritual dimensions at all and tend to reject reflection even on its possibilities in human cultural and intellectual history, and those who are "true believers" or firm adherents to a particular faith tradition and refuse to consider suggestions that the divine Spirit can be solicitous of intelligent beings elsewhere in the cosmos, to the extent of becoming present among them in some form appropriate to their context. They believe that even on Earth any faith tradition other than their own is at best a "heresy" (in the same religious tradition, but a doctrinal deviation from dogmas professed by the "true religion") and at worst a cause for coercive conversion, and in both instances deserving damnation in a life to come. They believe that only Jesus can "save" someone for "eternal life," and so even ETI, whatever their intellectual level and spiritual understandings (if any), must be baptized into Christianity—for their own benefit. Such a type of "evangelical" would have a particular zeal, assumedly, if humans found intelligent life that had evolved more recently than humans on a sister planet in space, and are more malleable into a Christian ideology. Ironically, then, the fundamentalist assumptions of reductionist science and religious dogmatism are equally challenged by the possibility of spiritual evolution for the same fundamental reason: they have come to absolutize and universalize their respective contradictory and conflictive perspectives.

Science, Spirituality, and ETI

In the early twenty-first century, inquisitive and spiritually aware (in terms of social perception, with or without a Spirit consciousness) scientists have mulled over the possibility of ETI existence and Earth region appearances. Among these biologists are molecular biologist Francisco J. Ayala, Professor of Biological Sciences, University of California-Irvine, and a recipient of both a National Medal of Science and the Templeton Prize; and entomologist Edward O. Wilson, University Research Professor Emeritus, Harvard University, and recipient of two Pulitzer Prizes. Ayala and Wilson have contrasting perspectives on religion/spirituality, similar views on "intelligent design," and somewhat contrasting views on the possible presence of ETI in the cosmos, where Ayala is more skeptical and Wilson more open.

Francisco Ayala's Scientific-Spiritual Perspective on Evolution and Intelligent Design

Francisco J. Ayala explores evolution and intelligent design from a scientific perspective principally in *Darwin's Gift to Science and Religion*. The title will startle readers who have been led to believe that there is an unbridgeable gap or insurmountable barrier separating science and spirituality (which in its institutional and social form is a "religion").

Ayala observes in "Darwin's Revolution: Design Without Designer" that in *The Origin of Species* (1859) Darwin called the evolution of organisms "common descent with modification," not "evolution." Ayala states that Darwin's principal project in *Origin* is to "account scientifically for the design of organisms." Darwin uses examples of humans' breeding cattle, dogs, plants, etc., to meet particular human purposes as a prologue to describing how other biota use a similar process for themselves in order to adapt to changing contexts. (One might say that biota in such circumstances effectively "breed themselves" to survive in new settings.) Ayala states that

> the design of organisms as they exist in nature ... is not "intelligent design," imposed by God as a Supreme Engineer or by humans; rather it is the result of a natural process of selection that promotes the adaptation of organisms to their environments. This is how natural selection works: Individuals that have beneficial variations, that is, variations that improve their probability of survival and reproduction, leave more descendants than individuals of the same species that have less beneficial variations. The beneficial variations will consequently increase in frequency over the generations; less beneficial or harmful variations will be eliminated from the species. Eventually, all individuals of the species will have the beneficial features; new features will continue accumulating over eons of time.

When closing this discussion, Ayala cites Darwin's thoughtful and appreciative words at the conclusion of *Origins* when he was thinking about the natural process of natural selection: "from so simple a beginning *endless forms most beautiful and most wonderful* have been, and are being, evolved."

Ayala notes, too, that a concept of "evolution," without using the term, was expressed by early Christian writers. Gregory of Nyssa (335–94) taught that the world emerged in two stages: first, by an instantaneous creative act of God, and second, through a gradual formative process over time. Augustine (354–430) taught that God created many species not directly but indirectly, in a seminal form that had the potentiality to exist materially when natural processes in creation gradually brought about their existence.

In "Arguing for Evolution," an excellent presentation of the meanings of "natural selection" and "evolution," Ayala states that "gaps of knowledge in the evolutionary history of living organisms no longer exist. This statement will come as a surprise to those who have heard again and again about 'missing links,' about the absence of fossil intermediates between reptiles and birds or between fish and tetrapods, and about the 'Cambrian Explosion.'" He continues: "molecular biology has made it possible to reconstruct the 'universal tree of life,' the continuity of succession from the original forms of life, ancestral to all living organisms, to every species now living on Earth. The main branches of the tree of life have been reconstructed on the whole and in great detail." Examples of such scientific discoveries, made in the nineteenth and twentieth centuries, include *Archaeopteryx* (discovered in 1861), the intermediate stage between reptiles and birds; and *Tiktaalik* (discovered in 2006), the link "between fishes and tetrapods" (four-limbed animals).

In his discussion of so-called intelligent design, Ayala states, "For a modern biologist who knows about the world of life, the design of organisms is not compatible with special action by the omniscient and omnipotent God of Judaism, Christianity, and Islam." After discussing the theological problem of trying to explain evil in the world when God is understood to be loving, all-knowing, and all-powerful, he notes that John Haught (Georgetown University Catholic theologian) writes about "Darwin's gift to theology," and Arthur Peacocke (Oxford University scientist and Anglican theologian and priest) called Darwin a "disguised friend" of theology—their shared conviction: theology has had a difficult time explaining *theodicy*, the problem of why, if God is loving, omniscient, and omnipotent, human evil and catastrophic natural events exist in the world—and God apparently does nothing about them. Haught, Peacocke, and Ayala all agree that evolution's unfinished natural processes cause pain, suffering, and death; these do not occur because of God's flawed design of or indifference toward biota. In his concluding paragraph on intelligent design (ID), Ayala declares, "Proponents of ID would do well to acknowledge Darwin's revolution and accept natural selection as the process that accounts for the design of organisms, as well as for the dysfunctions, oddities, cruelties, and sadism that pervade the world of life. Attributing these to specific agency by the Creator amounts to blasphemy. Proponents and followers of ID are surely well-meaning people who do not intend such blasphemy, but this is how matters appear to a biologist concerned that God not be slandered with the imputation of incompetent design."

On the relationship between religion and science, Ayala comments, "We will see that scientific knowledge and religious belief need not be in contradiction. If they are correctly assessed, they *cannot* be in contradiction, because science and religion concern non-overlapping realms of knowledge. It is only when assertions are made beyond their legitimate boundaries that evolutionary theory and religious belief appear to be antithetical." He quotes a National Academy of Sciences statement on the discussion: "Religion and science answer different questions about the world. Whether there is a purpose to the universe or a purpose for human existence are not questions for science.... Consequently, many people, including many scientists, hold strong religious beliefs and simultaneously accept the occurrence of evolution."

Regarding creationist assertions that Genesis takes precedence over science, Ayala quotes Augustine once again, this time the theologian's fourth-century statement about incompatibility between Genesis and science: "If it happens that the authority of sacred Scripture is set in opposition to clear and certain reasoning, this must mean that the person who interprets Scripture does not understand it correctly." Ayala elaborates thus: "It is possible to believe that God created the world while also accepting that the planets, mountains, plants, and animals came about, after the initial creation, by natural processes." He states further that while "science is a way of knowing," it "is not the only way.... Knowledge also derives from other sources, such as common sense, artistic and religious experience, and philosophical reflection." He is astonished when some scientists and others assert that "there is no valid knowledge outside science." He notes that while modern, empirical science began in the sixteenth century, for millennia previously people had built cities and roads, devised political and judicial systems, developed art, music, literature and agriculture, and engaged in philosophical discourse. He concludes his discussion here by stating that "successful as it is, and universally encompassing as its subject is, a scientific view of the world is hopelessly incomplete. Matters of value and meaning are outside science's scope."

Francisco Ayala affirms natural selection and evolution, but rejects "intelligent design" because of its biological inaccuracies and its "slander" against God. He affirms, too, religious explorations of the value and meaning of life. His perspective integrates well the complementary perspectives of science and religion.

E. O. Wilson's Scientific Reductionist Perspective on Evolution and Intelligent Design

In *The Social Conquest of Earth* (2012), Harvard biologist emeritus Edward O. Wilson elaborates at length his recently developed theory that evolution by natural selection occurs by multilevel selection, not by kinship selection and bias as had been widely accepted previously by Wilson himself and other evolutionary biologists (a theory related to Richard Dawkins' idea that a "selfish gene" controls human evolution and conduct in context). Wilson advocates now a "new theory of eusocial evolution." He declares that humans' evolution is "driven by both individual and group selection." This "multi-level selection consists of the interaction between forces of selection that target traits of individual members and other forces of selection that target traits of the group as a whole. The new theory is meant to replace the traditional theory based on pedigree kinship or some comparable measure of genetic relatedness."

In an earlier work, *The Creation: An Appeal to Save Life on Earth* (2006), for which this writer wrote an endorsement, Wilson argued, regarding science-religion relationships, for rapprochement between religion and science not on philosophical terms, but for practical, collaborative efforts to save life on Earth. Wilson grew up in the Southern Baptist tradition and wrote *The Creation* as an invitation to a representative pastor from his religious heritage to "meet on the near side of metaphysics" so that they might cooperate to save Earth, as "religion and science, working together, are the best hope" that Earth has. He urges Christians to act in accord with their religious faith's appreciation for and teachings on humans' responsibilities for God's creation, which would complement his secular humanist perspective and actions to save life on Earth. As discussed below, in his current work Wilson seems to have backtracked somewhat from this positive assessment of the relationship of science and religion.

Ayala and Wilson on ETI

Francisco Ayala and E. O. Wilson both consider the likelihood that extraterrestrial intelligent life exists in the cosmos. In the endorsement Ayala wrote for *Cosmic Commons* (2013), he states, "Whether or not you believe in the likelihood of ETI, and *I do not*, you'll find *Cosmic Commons* fascinating" (emphasis added). He does not "believe" that ETI exists but, as implied in his endorsement, recognizes that it is possible that people might find discussion on this subject thought-provoking and, for those who accept that ETI does exist, affirming. The fact that despite his current disbelief in the existence of ETI Ayala is willing to write an endorsement for the book might indicate that he is open to presentation of scientific evidence that might change his view. This perspective contrasts with "official" government and academic institutions' statements and policies, as described earlier. In contrast to government PR reps who speak confidently about UFOs and ETI without scientific credentials, he is a gifted international scientist who is open to independent and objective scientific exploration of the topic of ETI.

Edward O. Wilson in *Social Conquest* several times uses ETI to illustrate the evolution and complexification of Earth's biota. He muses early on, "Had extraterrestrial scientists put down on Earth three million years ago" the activities of ants, termites, and honeybees would have amazed them. Searching for species that had begun to evolve into intelligent beings, however, they would have concluded that the "African australopiths, rare bipedal primate species," were not promising as candidates for complexification. In fact, ETI likely would have concluded that probably no species would evolve to have intelligence for thousands of millennia. However, after ETI left to continue their scientific expedition elsewhere in the cosmos, "The brain of one of the australopiths began to grow rapidly." Over the next three million years, their species evolved into *homo sapiens*. If descendants of the earlier ETI scientists were to return to Earth today, they would be greatly surprised to learn not only that australopiths had survived and evolved into an intelligent species, but also that this descendant species was "destroying its own biosphere." They would find this very disturbing, and have little hope for the future of the species—or its world. Wilson comments, "The origin of modern humanity was a stroke of luck—good for our species for a while, bad for most of the rest of life forever." His careful tracing of the biological origins of intelligent life is interrupted periodically by his overall, ongoing concern about the future of life on Earth—and of Earth's future as it experiences the devastation made possible by the irresponsible, planet-altering and biota-extincting uses of human technology.

In his discussion of human traits Wilson notes that sixty-seven traits are present in hundreds of human societies described in the Human Relations Area Files (1945). He states that these traits might not be universal in the cosmos: "It is tempting to suppose that this list is not only truly diagnostic for human beings but inevitable for the *evolution of any species in any star system that reaches the human level of high intelligence and complex language*, regardless of its undergirding hereditary predispositions. However, that is almost certainly not the case, because it is possible to imagine other worlds in which large terrestrial creatures evolve different combinations of cultural traits. It would be premature to expect each of such theoretical universals to be genetic in nature" (emphasis added).

At the end of *Social Conquest* Wilson notes several times that ETI might exist, leaving it an open question. He speaks of humanity as "far and away life's greatest achievement," and adds, "We are the mind of the biosphere, the solar system, and—*who can say?*—perhaps the galaxy" (emphasis added), indicating that ETI existence or nonexistence has not yet been established. Later he states, complementing a central theme of *Cosmic Commons*, that we need to change our consciousness and conduct on Earth and take care of our home planet before we colonize other planets, so we don't act similarly elsewhere: "It is an especially dangerous delusion if we see emigration into space as a solution to be taken when we have used up this planet."

Wilson states that people should

> ask seriously why, during the 3.5 billion-year history of the biosphere, our planet has never been visited by extraterrestrials. (Except perhaps in fuzzy UFO lights in the sky and bedroom visitors during waking nightmares.) And, why has SETI, after searching the galaxy for years, never received a message from outer space? The theoretical possibility of such a contact exists and should be continued. . . . Of course, a scenario to explain the absence of extraterrestrials is that we are unique in all the galaxy going back through all those billions of years; and that we alone became capable of space travel, and so the Milky Way now awaits our conquest. That scenario *is highly unlikely* (emphasis added). . . . I favor another possibility. Perhaps the extraterrestrials just grew up. Perhaps they found out that the immense problems of their evolving civilizations could not be solved by competition among religious faiths, or ideologies, or warrior nations. They discovered that great problems demand great solutions, rationally achieved by cooperation among whatever factions divided them. If they

accomplished that much, they would have realized that there was no need to colonize other star systems. It would be enough to settle down and explore the limitless possibilities for fulfillment on the home planet.

In contrast to what Wilson states, there has been mounting evidence of an ETI presence. It is found in much more than "fuzzy UFO lights" and "bedroom visitors." However, there is considerable agreement with him in the scholarly community and among the general public that the scenario that humans are "unique" as cosmic intelligent life is highly unlikely.

Wilson's ideas here coincide with those of many scholars, particularly in the scientific community. Wilson and a majority of theorists, it seems, differ markedly from Stephen Hawking, who proposed that since we humans are destroying Earth we should send a representative group of us into space to start anew—as if somehow we won't do the same elsewhere if our current consciousness and conduct remain part of who we are as a species. This is especially true, it should be added, if we continue to be coerced or hoodwinked by the perspectives, practices, and power of members of our species who orient our economics and politics to benefit exclusively or primarily themselves and their socioeconomic class.

Wilson concludes *Social Conquest* with hope for humans on Earth: "So, now I will confess my own blind faith. Earth, by the twenty-second century, can be turned, if we so wish, into a permanent paradise for human beings, or at least the strong beginnings of one. . . . Out of an ethic of simple decency to one another, the unrelenting application of reason, and acceptance of what we truly are, our dreams will finally come home to stay."

Edward O. Wilson was one of the scientists to whom I dedicated *Cosmic Commons: Spirit, Science, and Space*—not because I agree with all of his ideas, theories, and statements but because I respect his exceptional intellect and commitment to ecological wellbeing for humankind, all biota, and abiotic Earth. These are expressed in his understandings of material reality (to him, the only reality), and consequently in his scientific creativity, research, field work, explorations, writings, and corpus as a whole. His deep sense of material existence complements my own, and also complements, by contrast, my deep experience and sense of the spiritual dimensions of existence and reality. His ETI ideas and perceptions overlap his religious ideas and perceptions: he requires material, science-based evidence that the material existence of ETI is credible, particularly when it is not seen, experienced, or intuited personally.

Science and Spirituality

The contrasting perspectives that E. O. Wilson and theists have in their respective views of spirituality can be illustrated by brief responses to select passages of *Social Conquest*. Wilson says that "religion will never solve" the "great riddle" comprised of the questions "Where do we come from? What are we? Where are we going?" which make us "terribly confused by the mere fact of our existence." Does religion really try to "solve" this riddle (a hopeless task for theist and atheist alike)? Ideally, religion (and more evidently, spirituality) seeks deeper, sacred dimensions of reality that would complement human materiality and morality and give humans a greater sense of our place—as much as we can imagine it, in different moments of time—and a holistic, multidimensional experience of the integral being of the cosmos. While *religion*, an institutionalized form of spirituality, would have some difficulty—because of its structural, culturally limiting and limited doctrines—addressing the "riddle," *spirituality*, as the fundamental and universal engagement with the Spirit, will be freer to suggest open-minded, open-ended, and in-depth consideration of this "riddle."

Wilson states that religious "myths," in his view, "live in the heart, not in the rational mind." He expresses a narrow understanding of what a "myth" means and how it is operative—and amendable, over time (the meaning and types of myth are explored in depth in *Cosmic Commons*)—in religious thought and practice. Myths seem to be for him fictional stories to express religious sentiments and beliefs; they are the "core of organized religion." He states that science and religion are in opposition, and their worldviews "cannot be reconciled" since they represent, respectively, "*trust* in empiricism and *belief* in the supernatural" (emphasis added). Why not belief in empiricism and trust in the supernatural or, even more to the point, *trust* in empiricism and *experience* of the supernatural? Empiricists believe that theirs is the only reality, while people with a spiritual consciousness accept and have faith in *both* realities: they cannot do otherwise, as material beings who explore spiritual realities in order to have the "whole picture" of humans in the integral being of cosmic reality.

Wilson declares that "the human condition is an endemic turmoil rooted in the evolutionary processes that created us. The worst in our nature coexists with the best, and so it will ever be." While at the present stage of our human evolution we might have such turmoil, it does not follow empirically that the worst and best in us "will ever be" in turmoil. Experiencing the material-spiritual dimension of the cosmos (or even the hope that this dimension exists) might well be a hint that this is part of a progressive evolutionary transformation of humanity: humankind might be becoming, through evolution, *more* rather than *less* spiritual, entering a stage of evolution where institutionalized religion is no longer "needed" or even "helpful" for spiritual experiences that influence positively material reality. Spiritual transcendence-immanence already is rooted to a greater or lesser extent in individuals and groups. It is a complementary aspect of human evolution that *might*, if it does not become primarily selfish, self-interested, or self-absorbed, enable humans as individuals and groups to emerge through evolution over time to mitigate, or even eliminate to a great extent, the "worst in our nature."

Spirituality is not confined to "out-of-body experiences." Spirituality, because of its openness to multiple dimensions of existence, is a fuller immersion in the depth and meaning of the materiality that defines us as human beings-in-the-world. As a species we are ever being-becoming through evolution and—if we do not destroy ourselves, all biota, and our Earth home—we might evolve to a "higher stage" of being where our spirituality is enhanced and we have a greater consciousness of our changing "place" in the integral being of the cosmos as a whole. In this evolutionary perspective, spirituality continues to evolve and extend, and might contribute to attaining ever higher evolutionary stages. (It is possible the ETI already has evolved to reach that stage of spirituality, or at least incorporated its elements into its culture, group dynamics, and individual and group behavior.) This spiritual stage would pass beyond ideas and dogmas about individual salvation and institutional religions' control of spiritual beliefs and practices.

Scientific thought and religious thought are not *competing* fields of knowledge, as Wilson believes. Rather, they complementarily wonder about, are interested in, and are curious concerning different dimensions of the cosmic existence of humans and other intelligent life. Science explores materiality; spirituality accepts human materiality and explores its spiritual meaning (including through reflection on materiality and, to some extent, mortality; it reflects more on the "why" behind existence, and the "who" that seems to be engaged as a transcendent-immanent Presence). Science helps us understand to an ever-greater extent our evolving material existence. Religion, and even more a universal spirituality, helps us understand our materiality and experience moments of our transcendence-immanence.

Spiritual evolution effects an increasing spiritual awareness that, from its base in materiality, transcends cultures and religions.

Wilson suggests that war is universal and eternal; it is not "an aberration of history, a result of the growing pains of our species." Perhaps not. However, we have, as ideals and in law, banned some particularly horrific weapons of war, rejected torture as acceptable, and proposed universal human rights and a quest for peace; these human strivings are embodied in the actions and documents of the United Nations. Might we be a tad more optimistic? Perhaps the spiritual evolution will help us, in a coming evolutionary stage, overcome the instincts and social behavior that prompted wars during our "growing pains" as we went from intellectual and spiritual infancy to adolescence. We can be hopeful that as we become more spiritual we will mature into adulthood as a species. In the spiritual evolution, people might meet more often on the "far side of metaphysics" rather than on the "near side of metaphysics," and then come to have a "shared metaphysics" over time.

Wilson declares that "Saint John the Divine might have experienced a real divine visit just as he reported it. Far more likely, however, he had dreams from taking hallucinogenic drugs." Why is his hypothesis "far more likely" than others, including what might be discerned in the experiences and teachings of spiritual leaders in different cultures at different periods of time? Wilson's assertion illustrates metaphysical speculation that flows from an intellectual (or even, possibly, psychological) projection. Apparently, even though personally he has not had moments of transcendant spiritual experience and professionally has not done research into the topic, he still feels comfortable enough in the area to project and superimpose his materialism onto metaphysical narratives. Since he has not had spiritual experiences or engaged with those who have, he believes he can dismiss them: they cannot be true or "real" because he has decided *a priori* as a universal absolute that the spiritual realm does not exist. He will try to explain away what he does not comprehend, having no desire—or time, given his ongoing scientific work—to investigate reports or expressions of spiritual experiences seriously, as he would do regarding matters in his field of scientific research.

Perhaps ironically but complementarily, while Wilson imposes material understandings onto religious metaphysics, creationists impose religious understandings onto material realities described by science. Each strays beyond their particular "area of expertise" and becomes an alien invader of the other's intellectual territory.

Wilson sees contemporary religions as an outgrowth of tribal creation myths, and has a strong belief that religions are focused on control, requiring "submission to God.... Yet let us ask frankly, to whom is such obeisance really directed? Is it to an entity that may have no meaning within reach of the human mind—or may not even exist? Yes, perhaps it really is to God. But perhaps it is to no more than a tribe united by a creation myth. If the latter, religious faith is better interpreted as an unseen trap unavoidable during the biological history of our species. And if this is correct, surely there exist ways to find spiritual fulfillment without surrender and enslavement. Humankind deserves better." Indeed we do. We find "better" already in noninstitutionalized spirituality, and in religious commitments to social justice and environmental wellbeing for the good of people and Earth. Not all peoples of religious faith seek to reinforce obeisance to their leaders, doctrines, and structures internally, and impose them on others externally because they have a different (theistic or atheistic) faith.

Wilson's statement on evolution—"all of life's *material* entities and processes have arisen through evolution by natural selection"—requires, it would seem, a complementary clarification: "*... and intelligent life's spiritual understandings historically have evolved and progressed when an ongoing attitude of openness, discernment, and engagement, during the time homo sapiens has been evolving materially, has been present and operative.*" Spiritual evolution can continue during and complement biological evolution and social evolution, and might become integrated with both.

Theists should not be offended by, nor attack Wilson for, his beliefs and their expression, but instead understand that what he proposes is his honest and direct presentation of his worldview as a humanist scientist, and his frustration with religions' rejection rather than reception of their responsibilities toward—and from within—what they view as God's creation.

People with a theistic perspective should be neither offended nor surprised by what Wilson writes. They should instead understand that he is an excellent and eloquent representative of secular humanist thought, and his ideas can help theists to clarify and sharpen their thinking. Theists need not be angered, as some have been, by what they perceive to be an attack against their beliefs; rather, they might regard his strong statements as an affirmation of his beliefs. He does stray into metaphysics on occasion, despite being an atheist materialist. His arguments have no firm foundation in materiality, a different dimension of reality than spirituality, he subjectively uses objective criteria from materiality to judge metaphysical and spiritual understandings. Just as metaphysics cannot judge the accuracy of science claims, lacking the tools to do so, materialist tools cannot judge the "truthfulness" of religious claims, lacking the knowledge to do so.

Wilson and many theists would be in agreement about the fact that religious leaders on occasion do use their power and authority over their adherents, and seek to extend it over others—theists and atheists—who do not share their ideology, for covert political and economic purposes, to obtain benefits for themselves and their religious institution. Creationism and intelligent design are two of the weaker forms of religious imperialism (they lack a majority national base for political power, and sufficient funds to overcome in society other churches and atheists in order to extend their economic power). In recent years in the U.S., too, Catholic bishops attacked the Affordable Care Act (ACA) and President Barack Obama, falsely stating that "Obamacare" was an assault on "freedom of religion" because it required that employees in nonreligious work but employed by church bodies must have birth control in their health plan. The bishops sought to impose official Catholic teaching about birth control (a position that is not accepted by the vast majority of Catholics) upon women in their employ (whether or not Catholic) who sought to have personal control over what should be *their* voluntary decisions to exercise responsible procreation practices, and express their love sexually. Right-wing groups quickly supported this tactic, hoping to further their ideology. Despite the fact that the ACA embodied many Catholic social principles, rarely did bishops as a whole or in part publicly express support for provisions of the act that were congruent with Catholic Social Teaching—as was the case with the vast majority of the health policies and practices in the act. Conservative Catholic leaders' practices here mirrored conservative Protestant leaders' tactics to force the teaching of the evolution-denying religious doctrine of ID in public schools, and to restrict efforts to curb global climate change that is harming people, extincting species, and threatening Earth's survival.

Although examples of religions' overextension into the political sphere are certainly extant, including in religion-backed social conflicts and wars, this over fusion of religiosity and politics should not be used to deny that core religious ideals also have had positive social impacts—as evidenced by, among others, Catholic bishops' strong support for social justice, living wages for workers, racial integration, world peace, and environmental wellbeing, and Methodist leaders' and laity's support for the abolition of slavery, women's suffrage, social justice, and environmental wellbeing.

Spiritual Evolution

Consideration of a corresponding, complementary form of human evolution would supplement in a significant way E. O. Wilson's discussion of human evolution. An ongoing "spiritual evolution" in human thought seems to be underway—an increasing spiritual awareness that transcends human cultures and religious institutions and will be the next evolutionary stage in human consciousness. It might well help overcome human instinctual behavior that becomes adverse social behavior (such as conflicts that lead to wars among members of humankind). A spiritual evolution might have occurred, too, in the past and present evolution of ETI. In terms of TI evolution, Wilson states that chimpanzee and human territorial aggression might have been "inherited . . . from a common ancestor or . . . evolved . . . independently in response to parallel pressures of natural selection and opportunities encountered in the African homeland." He theorizes that it is "more likely" that aggression occurs because it is a trait that was received "through common ancestry." However, it could have evolved due to the contextual influences he describes. If the former is true, it might be an unalterable trait; if the latter is true, territorial aggression is not a genetic "given" for humans nor, in cosmic contexts, for ETI that underwent a comparable evolutionary development.

It is possible, in the latter case, that ETI either (*a*) never had this trait in their planetary context, or (*b*) has evolved—culturally and/or biologically—beyond it. Similarly, regarding spirituality and religion, ETI might not have "needed" it as an evolutionary adaptive trait, but developed it intuitively or because of transcending experiences; it might still be evolving today, well beyond institutional religions—which themselves might never have emerged in ETI's planetary contexts and evolutionary complexification and social progression. Spirituality, in this event, is an ongoing evolutionary process, a coming (for humankind) higher phase of evolution, a spiritual awareness that is beyond cultural diversity and religious divisions and institutions. In evolutionary terms, too, spiritual transcendence-immanence, already rooted to a greater or lesser extent in individuals and groups, is a complementary aspect of human existence and evolution that could ameliorate to a significant extent the "worst in our nature." Spiritual evolution, like biological evolution, continues to develop and to be enhanced and extended as an aspect of human consciousness. It could provide an evolutionary contribution to and advantage for humankind's survival and wellbeing, and become embedded in human consciousness as part of humankind's integrated material-spiritual evolutionary stage.

Contemporary biologists Francisco J. Ayala and E. O. Wilson have contrasting views on religion and ETI: while Ayala is open to spiritual dimensions of reality, Wilson is not; and while Wilson is open to ETI existence, Ayala is not—at this time. In terms of "belief" in the unseen, both scientists await material evidence: Wilson indicates that there is a possibility that ETI exists, while Ayala suggests that a divine Spirit exists.

Secular Humanism and Spiritual Humanism

Wilson discusses the evolution of religion from his perspective as a *secular humanist* (a self-characterization he expresses in *The Creation: An Appeal to Save Life on Earth*). An alternative, *spiritual humanist* perspective, by contrast, incorporates knowledge from science—the materiality present in cosmic existence and existents, which humankind discovers over time and expresses in scientific data—and reflections that emerge from a sense or experience of the Spirit. Spirituality is not to be confused with religion or theology. Distinctions among them follow, in a very simplified (and, hopefully, not too simplistic) core summary of each. ("Materiality" is a generic term, in these considerations, for space-time influenced, intertwined, and interchangeable matter-energy.)

Spirituality is the most personal and most universal engagement with the Spirit. It is personal in that it is an individual's unique experience; it is universal in that all unique personal experiences are distinct but related engagements with one Spirit—the transcosmos universal Spirit who is immanent (inherent in all being) and immanent-transcending (present in creation as a personal entity but not defined or limited by creation), and is Being-Becoming (having both core timeless attributes and ongoing absorption and internalization of creatures' experiences in time—such as joy and sadness, pleasure and suffering—as biotic evolution occurs on Earth and in other cosmic places and leads to TI and ETI).

Religion is the institutionalization and socialization of spirituality. Religion provides a structure, doctrines, and rituals through which exceptional and highly regarded individuals' spiritual experiences and related ideas and teachings are conserved, disseminated, and, ideally, practiced in a way imagined and intended by the intentional or unintentional founder of the religion. People adhering to a particular religious tradition strive to replicate over time what the earliest practitioners received in the religions' origin moments. Such institutions can become rigid and self-serving over time. The social cohesion of distinct faith communities can, by contrast, be mutually reinforcing for faith and encourage compassionate works of love and justice and thereby have beneficial societal impacts beyond the specific religious community from which they emerge and by which they are expressed.

Theology is the intellectualization and rationalization of religion. It ventures beyond the spirituality experienced and expressed by the revered teacher who engages the Spirit, an individual's experiences of the Spirit, and a religious institution's teachings and practices that compound as time distances it from its perceived founder. Theology is two steps beyond the initial spirituality of both the "founder" and of individual spiritual persons in subsequent times, places, and cultures. Its internal institutional discussion and debates over doctrines can stimulate thoughtful, in-depth consideration of a religion's core and corollary beliefs and practices.

Spirituality, then, is the most fundamental and important of the three manifestations of theistic faith: it is both personal and universal, and ultimately can bridge ideological and doctrinal chasms and integrate communities and individuals. Religions go to war over unverifiable dogmas, and theological disputes rage over unprovable doctrines; but spirituality as an experience and expression of the Spirit can effect harmony in community. The contrasting and yet complementary perspectives of secular humanism and spiritual humanism can be integrated to promote care for creation and creatures. As noted earlier, Wilson himself declares in *The Creation* that Christians and humanists should "meet on the near side of metaphysics" because religion and science, working together, provide the best possibilities to save Earth's biota from extinction.

Wilson told me several years ago that evangelical Christians, after the publication of *The Creation*, invited him to major conferences, sometimes with more than a thousand evangelicals present, because "they appreciated it that a scientist was reaching out his hand to them." I hope that Wilson, evangelicals, and representatives of each of their respective faith traditions, atheist and theist, secular humanist and spiritual humanist, will continue to "reach out a hand." They will express, then, instead of a particular absolutized perspective, their recognition of and openness to their complementarity. In this way they will be able to "meet on the near side of metaphysics" and work on common ground in common efforts to promote the common good of (the) Creation. This perspective and practice will serve humanity well, too, should extensive Contact occur and interpersonal TI-ETI engagement and exchanges of ideas be shared.

Wilson does not address specifically the concept of intelligent design. It has no relevance for him because as a secular humanist he is tracing the biological and social origins of religion, not current religious beliefs *per se*. Obviously, however, he and Ayala share a common opposition to ID's science-rejecting, evolution-opposing ideology.

TI and ETI Worldviews and Worldsviews: Atheist, Theist, and Alien Faiths

Currently and concurrently, diverse cultural and ideological groups and individuals propose worldviews that developed from their specific historical context. Some groups invite others to consider their particular perspective and can stimulate mutually profitable exchanges of beliefs and ideas; other groups strive to subject others to their way of thinking, thereby catalyzing conflict. On Earth, TI atheist and theist faiths conflict; in the cosmos beyond, the types of ETI faiths, if any, are as yet unknown to humankind. In all cases, as we recall the meaning of *alien*, the faith expressed by others may be regarded as "alien": foreign to our consciousness, culture, or context. Atheist and theist faiths are in this sense alien (unknown to and unaccepted by the "other"); TI atheist and theist faiths are initially alien to ETI, just as ETI beliefs initially are alien to TI.

In terms of TI-ETI faiths' interaction, fundamentalist Christians' religious resistance to accepting ETI spirituality or atheism has two distinct TI adversaries (ETI might be presumed to be an alien adversary when interacting with *fundamentalist Christians*, since ETI would have their own evolutionarily developed understanding of faith; and fundamentalist Christians would have an alien adversary when interacting with *fundamentalist atheists*, if ETI were to have had a spiritual evolution that led to a higher level of spiritual—albeit not religious—consciousness).

First, among human beings globally and even in fundamentalist Christians' home nation, the U.S., there is relatively little societal or even individual acceptance of fundamentalists' doctrine that God became part of the cosmos only on Earth, in Christ Jesus, to "save the world"; that Christianity is the "true religion"; that the Bible is the only or the preeminent sacred book on Earth; and that the older, sixteenth-century King James translation of the Bible is the only "true" or "real" translation (any translation from one language to another, from diverse cultures into one, and from several time periods into one, embodies interpretation as words are selected by the translator). Even among Protestants, King James probably is used less than the combined uses of other translations in most denominations, and is not used at all by Catholics, who represent approximately 25 percent of the U.S. population. The King James Version is written in a poetically pleasing flowery English no longer used in English-speaking countries and has been rendered historically inaccurate in parts because of two centuries of archaeological discoveries—ancient ruins, battlefields, historical records, previously unavailable manuscripts of biblical books or segments thereof—which have enabled scholars and translators to provide more historically and linguistically accurate versions; this scientific data did not exist at the time any of the biblical books emerged as either oral traditions or writings (people no longer believe that the sun, planets, and entire universe revolve around Earth) were written. Further, the King James has outdated vocabulary, including words or meanings of words no longer current, some of whose dogmatic understandings, statements, and assertions are no longer accepted in Christianity itself, let alone by theists external to the Christian church. The fundamentalists have determined, despite all of this, that for their own eternal good and earthly wellbeing every other intelligent being encountered in the cosmos has to be baptized to become their version of a "Christian" in order to be "saved." (Fundamentalists conflict with other Christians, too, when they try to impose in public schools and on public places a poster or monument with the King James version of the Ten Commandments. Based on the same biblical texts, and alternative translations of their words, Jews, Catholics, and Protestants have three different versions of how these commandments are to be numbered.)

Second, primarily among atheists—whether or not fundamentalist nonbelievers in *Theos*, who have faith and security in their own beliefs—or secular humanists, religion itself, including Christianity, is a biological and social construct that evolved in human history. As with biblical fundamentalists, atheist fundamentalists start from a position of faith—for the former, that God exists; for the latter, that God does not exist—that cannot be empirically verified by either ideology. Science has to do with the observable, quantifiable, falsifiable materiality of being; spirituality or religion has to do with the materiality of being both as is, and as considered in relation to a nonquantifiable and nonfalsifiable nonmateriality of being.

"Faith" and its expressions can be neither "proved" nor "disproved." This is true for both atheist faith and religious faith. A "true *believer*" (when this concept is expressed in a positive and literal sense) is one who believes and does not require rational arguments for their beliefs, which are based in and oriented toward a spiritual, not rational or material, dimension of reality. They remain unshaken in faith because material-based humanist theories and (perhaps competing) religions' teachings ordinarily have little or no impact (unless they come to provide an accepted complement to—and sometimes a replacement for—elements of the believer's initial religious faith tradition).

The Christian Scriptures are specific about the meaning of *faith*: "Now faith is the assurance of things hoped for, the conviction of things *not seen*" (Heb 11:1; emphasis added). Note the words "not seen": this provides a clear distinction between material existence and the spiritual reality about which Christians have the conviction (not the knowledge) that it exists; materiality and spirituality come together in the Christian, who is living as a material being and is not a disembodied spirit. Christians and other theists see, physically experience through their senses, and *know* about the material world in which life is lived, but they *believe* in a spiritual dimension of that world which they realize is unseen. Materiality and spirituality are distinct but not exclusive dimensions of reality. Science rightly researches into and collects data about material reality, and suggests hypotheses and theories about aspects of material reality not apparent or even not (yet) known to exist. (Once something is "known," of course, it moves into the realm of data and fact, and need not be something about which a hypothesis or theory is proposed, or in which a belief is expressed.)

Humans, as groups and individuals with diverse types of spiritual consciousness, curiosity, and community contexts (including as distinct religions or denominations), rightly think and theorize about the spiritual dimension of reality. They understand that in reality their existence is multidimensional; they study the physical world and acquire knowledge about its material reality; they engage in "metaphysics" (thinking in a way that transcends or goes beyond physical, material existence) and formulate religious faith-developed doctrines, rituals, and codes of conduct that they believe enable them to participate in a spiritual reality; they identify their particular understanding of a spiritual reality and particular type of participation in a spiritual reality; and they faithfully follow the guidance and commandments of a Spirit or Spirits, as they understand them. They seek to reflect on what such a spiritual reality might be—sometimes by projecting into it their knowledge of material reality, including characteristics of the best human traits and conduct in material reality, and subsequently formulating characteristic aspects of or actions by sacred beings; and, at other times, by unexpectedly experiencing spiritual reality directly, with or without seeking it, as is noted about and expressed in the visions of spiritual leaders across religions, cultures, historical times, and specific geographic places.

The Christian Scriptures state, too, regarding Christian belief, that "we look not at what can be seen but at what cannot be seen; for what can be seen is temporary, but what cannot be seen is eternal" (2 Cor 4:18), and that "we walk by faith, not by sight" (2 Cor 5:7). Theistic faith, then, cannot be "explained" or "rationalized"; it might, however, be imparted by a Spirit; "doctrinized" and "ritualized"; and speculatively "theologized" and intellectually "elaborated."

In atheist faiths, all reality is material and can only be understood as entity embodiment of elements and energies that are cosmic existents.

In theist faiths, reality is matter and spirit intertwined. Matter is visible, experienced and assessed by the mind and the physical senses; tools are developed to enhance matter and enable it to be tangible and quantifiable. The spiritual is not-visible, experienced in transcendent-immanent moments, inaccessible to examination by methods used to assess matter. TI theists can come to understand in greater depth, in their interaction with ETI, the ancient Hebrew teaching in the Shema: "Hear, O Israel, the Lord your God is One." They become aware to a greater extent than before Contact that the One is viewed in different ways through the eyes of intelligent species from diverse times, places, and cultures, and their respective perceptions of the One are expressed in distinct ways in diverse tongues. Here again, their core spirituality enables them to accept and appreciate others' expressions of these others'—including aliens' and strangers'—strains of spirituality, which might be presented in specific doctrines and rituals.

Atheists and theists each have a core faith which cannot be "proved" to be true or false by either materialist contentions or metaphysical claims. In fact, for an atheist to make a "metaphysical" claim about a spiritual reality means that they are thinking about religion while in the realm of religion—this contradicts the essence of atheist belief, that there exists no "meta-" to the "physics" of material reality. (Philosophy *per se* is an exception, when atheists are exploring and debating ideas about the ultimate meaning of existence, if they believe or acknowledge that universal meaning exists, or at least a meaning which is workable for a time and place.)

One result of Contact might be a "conversion" to a different perspective, perhaps to one radically distinct from or contradictory to one that was regarded previously as unchangeable because of the assumed eternal "truths" upon which it was based. Conversion from tradition can occur as a result of some compelling intuition or innovative idea, or an unanticipated event or experience: a materialist might have a truly transcendent experience of a spiritual reality or Spirit Being; a spiritual person might encounter a compelling idea or insight from the material world, from observation and analysis, perhaps, or from another person or their written ideas.

For theists, evolution—including social evolution and religious evolution—can be viewed as a consequence of the creativity and freedom seeded by the Spirit into the cosmos in the cosmic singularity, the "day without yesterday," in the words of Georges Lemaître (who conceived what became known as the Big Bang theory), as an emanation of divine Being that gradually unfolds in the presence of divine love through eons of time. (Maximus describes a *Logos-logoi* dialogic relationship between the Creator and all components of creation, biotic and abiotic.) It becomes expressed with ever-greater autopoeitic diversity and complexity according to its ever-more creative capabilities that continually emerge in successive cosmic "moments." Just as biological evolution through core constituents of DNA enables life to adapt to changed environs when conditions emerge that require a new form or ability of a species, so too might it have some characteristics, or elements, genes, or DNA or RNA component that alone or in combination can catalyze emergence of a spiritual sensibility when the evolutionary moment is appropriate for this adaptation. The theist, to understand and accept this, must have accepted an appropriated and internalized evolutionary theory. Ultimate questions remain in cosmological and biological evolution, such as these: How did it all begin? What caused this beginning to begin? What, if anything, was present when it began? Was there *creatio ex nihilo* (creation from nothing), *creatio ex materia* (creation from existing matter), or *creatio ex Dei* (creation emanating from divine Being)?

Theists have faith that there is an immanent-transcendent Spirit in and beyond the visible cosmos; atheists believe that there is no Spirit in or beyond the cosmos. Each can help the other clarify their beliefs as they consider the other's thought.

Ecumenism on Earth and in the Heavens

A new Earth ecumenism would ensure that theists respect each other's faith, from every religious or spiritual tradition and from secular humanist thought; and that atheists respect each other's faith, and theist traditions. Christians, for example, can respect and, in that regard, accept as complementary to their own beliefs the materialist beliefs of E. O. Wilson and other atheists for whom materiality is the only reality—period. In the materialist belief system, there is no spirituality because such an evolved social construct is not based on (material) "reality." Conversely, for theists, materiality does not define the *entirety* of reality, only one of its aspects, and they can appreciate and respect atheists for providing insights, data, and theories about material realities. Similarly for Christians, the theistic beliefs of Judaism, Islam, Buddhism, Hinduism, and other religions, and the diverse spiritual beliefs expressed by indigenous traditions and in New Age spirituality, can be viewed as complementary perceptions of the spiritual dimensions of reality and of the Being who is immanent in all of them and in material reality.

Some scientists state that, by scientific criteria, Spirit, spirits, and spiritual dimensions are not real because they cannot be quantified or falsified. However, that is a biased definition of "real": those who believe that there is no nonmaterial existence want to subject beliefs in that existence and the very existence itself to tools and criteria previously used solely to assess data, hypotheses, and theories of materiality. Then, although by its nature a spiritual reality has no materiality of its own, and cannot be assessed by such criteria and tools, they want to use their criteria and material tools to examine spiritual existents. Somehow they decide then that their materiality-oriented tools and materially developed criteria are the only means acceptable to determine "reality." So, in sum, they believe that materiality is the only reality, and that with their materialist consciousness they can use the criteria and tools with which they examine, weigh, and judge materiality to provide, too, the intellectual and technological instruments they use to assess spirituality—which they believe does not exist. It follows from this convoluted reasoning, then, that spiritual being is nonexistent. Hmm.

Despite the foregoing, in terms of materiality itself and the exploration thereof, theists should appreciate the contributions of atheists to help them clarify their own thinking; to understand historical events interpreted theologically, and early religious dogmas expressed prescientifically; to distinguish religious themes and symbolism from science and history (the Bible is, after all, neither a science text nor a history book); to see the elements of myth in their narratives; to understand prescientific ways of imparting information, beliefs, and moral instruction (particularly through stories); to express religious beliefs and mores; to recognize the origins of their sacred texts at particular historical moments, eras, and places in specific cultures in a specific language, perhaps conveyed orally before they became written texts; and to gain through all of this a deeper spirituality and deeper understanding of both the material and spiritual past and present.

A sociocultural benefit of atheism is that it prompts some scientists, because they reject a "God hypothesis," to probe carefully and deeply into the history and essential aspects of material existence unrestricted by religious beliefs or taboos; they can use the data accumulated in this quest to persuade theists to reflect deeply on their beliefs. If atheists did so with respect and not condescension, theists would understand better the origins and potential continuing use of sacred texts, overcome current cultural constraints that are barriers to understanding these texts, avoid an assignation of "miracle" to unusual events, not impose their own beliefs on the meanings of biblical texts (the practice of *eisegesis* rather than *exegesis*), and not posit a "God of the gaps" as a creation-extrinsic cause of something currently viewed as "extraordinary" rather than "ordinary." Such previous beliefs came about because theistic believers did not recognize that they cannot understand something if they lack rudimentary knowledge of energy, materiality, geodynamics, biology, history, and social psychology, among other fields of inquiry.

Adherents of diverse religious traditions and atheist faith traditions should not intrude as alien invaders into each other's domain in order to judge, from their perspective, others' beliefs, stories, and statements, and engage in ideological warfare to prove their own ideology and disprove their adversary's reality.

Adherents of religious faith traditions and atheist faith traditions should strive to have an ecumenical spirit, expressed (I almost said "embodied") in a new Cosmos ecumenism: respect for intelligent beings from every faith tradition, atheist and theist, as expressed by their own kind and by other worlds' inhabitants; and openness to both the complementarity of their respective perspectives, and the shared understandings they have that might be foundations for collaborative projects to promote cosmic wellbeing and the wellbeing of their own origin worlds.

Let's get beyond the so-called war between science and religion; let's "meet on the near side of metaphysics," to borrow Wilson's apt phrase, and see if, as he suggests, science and religion can work together to save life on Earth without worrying about life in heaven. In this type of ecumenism, acknowledgment of others' faiths, including their understandings of divinity, might be one way of interpreting "Hear, O Israel: The Lord our God, the Lord is One" (Deut 6:4). God-Allah-Yahweh-Tunkashila is One, no matter the culture-originated names expressed in diverse languages by which Spirit is addressed in diverse religions. "Ecumenism" has meant, since the Second Vatican Council of the Catholic Church (1962–65), openness to and efforts to achieve a global unity among Christian churches, transcending diverse and divisive doctrines (thereby being a variation of "meeting on the near side of metaphysics," but sharing in common primary metaphysical values and most Christian doctrines, with some variations in dogma) to interact with respect, harmony, and mutual acceptance. A "new ecumenism" would extend this exclusively Christian effort outward toward unity characterized by a new inclusivity that invites fruitful exchange with all other faiths (theist and atheist). A "cosmos ecumenism" would be significantly more inclusive: it would invite TI and ETI to share their respective understandings, to the extent they exist, of spiritual dimensions of reality. As discussed previously, it might be the case that ETI evolved to a higher level of spiritual understanding or even a more spiritual existence, or no longer has (if they ever did) a sense of spiritual reality.

Philosopher George Santayana's (not to be confused with the popular singer, Santana) oft-quoted phrase is worth quoting in this context: "Those who cannot remember the past are condemned to repeat it." Religious wars once raged over finer points of doctrine, not really over belief in a deity but in undefinable characteristics of the deity, and "proper" ways in which, each proclaimed, the deity wished to be worshipped. Politico-economic wars are on the verge of once again breaking out—and religions might be enlisted in nationalistic ways to support one side or the other—as some humans eye greedily the lands, seas, surface natural goods, and subsurface natural goods in others' territories. The rush to exploit Arctic petroleum deep underwater, if it becomes a reality, is projected to have catastrophic environmental impacts both locally in the waters and globally in the air and on Earth's climate. The conflict over scarce and disappearing water in Africa, stricken by global heating's droughts, has ethnic and religious elements present already.

As TI who will journey Earth → exoEarth, what lessons will we have learned from our Earth context that we will take elsewhere as we venture outward to colonize? Will the U.N. Outer Space Treaty be strong enough to prevent greedy transnational/transstellar corporations from privatizing and exclusively profiting from exoEarth territories and the natural goods they contain? How will we humans, as aliens, interact with native beings?

As ETI journey exoEarth → Earth, what reception will we TI give to ETI, who might have evolved over billions or at least millions of years, progressed to greater complexity and intelligence—and perhaps to a different material manifestation, even one that is material-spiritual in some respect—and organized a communal culture that benefits all? If ETI has developed a common religion—or a common understanding and practice of a more advanced form of spirituality without religious institutions, clergy, and specialized rituals—economic wellbeing, and a consensual political system that equitably meets the needs of all, would humankind be willing to learn from and adapt to aliens with such perspectives and practices? Will Adam Smith's "invisible hand" reach greedily into space to grasp territory and natural goods on alien worlds without respecting others' cultures and economic practices, as has been done on Earth, justified by the Discovery Doctrine? Will evangelicals go forth to baptize ETI? Or will the complementary expressions of compassion in community evident in the writings of Thomas Paine, in diverse religious traditions, and in U.N. documents and ideals finally prevail?

The Spirit is One. This acknowledgment might be seen in a new light in a spirit of worldwide ecumenicity and *worldswide* ecumenicity such that it includes recognition that intelligent beings around the world and throughout the cosmos are worshipping the same sacred Being, whose manifestations and presence are recognized and experienced in diverse ways globally and cosmically. The One is the sacred Being whose transcendence-immanence is acknowledged universally, even worshipped cosmically. In cosmos contexts, this shared respect should be manifest in a spirit of cosmic ecumenicity, in which intelligent beings from all places in the cosmos would be viewed, too, as related to the same Spirit, the mind, heart, and soul of integral being.

Aliens to Aliens: Surpassing Institutional Impediments to Interrelationship

Humankind soon will venture into outer space personally, voyaging to distant places far beyond the moon we once walked, beyond Mars upon which only exploring machines have crawled. Once away from Earth, humans will be extraterrestrials everywhere they roam in the cosmos, possibly as TI aliens walking on ETI lands (actually, as ETI Earth aliens walking on exoEarth TI lands). On worlds previously unvisited, should TI and ETI arrive simultaneously or within a short time of each other, "alien" loses its divisive meaning to a certain extent: TI and ETI both are aliens to the new world and aliens to each other. When we recognize that TI and ETI are both "thinking stardust" that emerged in seminal form simultaneously on the "day without yesterday," at the moment of the Big Bang, we are enabled—and prompted—to seek our commonalities.

When we overcome what could divide us, and embrace what we have in common that could unite us, we will transcend government, military, and corporate propaganda that seeks to sharply distinguish among us; ethnic or species prejudice toward any of us; the academy, which refuses to acknowledge even the existence of a thinking "other"; religions that deem us doctrinally inferior; and a scientism that not only focuses exclusively on materiality, but tries to use tools useful for material research to assess spiritual understandings. When we have accomplished this social and spiritual transformation we will be enabled to meet on spatial, intellectual, material, and spiritual common ground.

6

ETI IN *AVATAR* AND *DISTRICT 9*

Aren't We Always the "Good Guys"?

IN THE TWENTIETH CENTURY, people became increasingly curious about the possibility that there might be intelligent life not solely on Earth but elsewhere in the vast universe. Although speculation about that possibility can be traced back to philosophers in ancient Greece, it was not something of particular interest to the common people. They lacked the leisure of philosophers; instead of casually considering and speculating on transcending topics they had to be focused on their everyday life, and the physical labor required for survival or at least for daily sustenance. Besides, they usually attributed extraordinary "signs" in the skies at night to specific acts of their deity or deities, who were putting them on notice that unusual events (good or bad) were about to occur, or that a particular person's birth would be significant, and so on. It never occurred to them, who were Earthbound, that beings somewhat like themselves would have the capabilities to traverse the heavens. The highest celestial realm, heaven, was the abode of the deities (their own and those of other peoples); mere mortals were not permitted to trespass, and those who did would be severely punished (a good example of this is the story of one of the Titans, the giant Prometheus, who stole fire from the gods on Mount Olympus and gave it to humans; Zeus, the chief god, ordered his punishment: he was chained to a rock, and every day an eagle flew in and ate his liver, which kept growing back—and the eagle kept coming back).

Millennia later, once humankind had developed the technology to fly, speculation began that intelligent beings somewhere beyond Earth (about whom ancient Greek philosophers had philosophized) had accomplished this too—and perhaps had developed it to the extent of being able to voyage great distances. Creative authors wrote science fiction novels and short stories. When these were published, the new genre became popular quickly and its stories were imitated or elaborated creatively; this stimulated the public imagination further.

Consequently, people in the U.S. began thinking seriously about extraterrestrial beings, wondering how extraterrestrials might impact Earth and humankind. They became increasingly concerned that extraterrestrials might not be friendly. Popular, often subconscious or suppressed speculation about ETI became especially evident in public reaction and near-irrational fear throughout the country when, in 1938, Orson Welles narrated and directed, in a live radio broadcast, an adaptation of H. G. Wells' *War of the Worlds*. People listening to a music program on the radio thought that the reading, which Wells artistically interjected in segments interrupting the music, was actually a series of breaking news stories about dramatic and frightening events. The news flashes described the landing and actions of invaders from the not-so-distant planet Mars, and humans' futile efforts to resist them. A mass panic ensued. People called government agencies, newspaper offices, and radio stations to find out the latest news and also to report on Martian invaders' ships that they had seen. This was remarkable conduct: U.S. citizens internalized and then personalized supposed current events, and participated in them by "seeing" related incidents. Apparently, the "news" narration brought to the surface people's long-held and perhaps intentionally suppressed fears regarding their expectations of space aliens' conduct.

It's reasonable for us to assume that a majority of U.S. citizens at the time had not read Wells's novel (published in 1898, forty years earlier) or, if they had, either did not make a connection between the novel and the radio broadcast, or did connect them and believed that Wells had been prescient in narrating a fictional alien attack that became an instant "fact" when presented as radio news. The number of people likely to have read the story is very few relative to the number of people who panicked at the "news" broadcast. People's overreaction to a voice on the radio prompts the question, Where did they get information and develop beliefs that intelligent beings exist extraterrestrially; inhabit Mars, a nearby planet; would invade

Earth; and would attack Earth's inhabitants? It's unlikely that this was an instant insight at the moment of the broadcast. Rather, it indicates that numerous people "believed in" a threat that they thought space aliens posed to people and planet, and thought that their worst fears and nightmares were being actualized in material reality. Perhaps the seeds of this understanding included a normal "fear of the unknown"; instinctual fear of a predator, or of representatives of an aggressive culture that was technologically and militarily superior; or the stimulus of previously published fictional narratives which suddenly seemed to have described what was now a reality, a nonfiction event. Several or all of these factors had prepped the populace for and imposed a psyche-disturbing meaning upon the "news" they heard.

Extraterrestrials and Aliens

The meanings and implications of "extraterrestrial" are more evident when we write "extra-terrestrial." It describes someone or something that is extra-terra: from beyond *terra*, that is, Earth as a whole; or, from beyond one's particular territory, the *terra* that is a community's or a culture's homeland: "they" are *extra* to "our" *terra*. The extent of the area of a people's territory might be delineated geographically by some type of formal boundary markers peoples place to indicate "borders"; by geophysical characteristics (such as mountains, hills, or unusual natural rock structures); by rivers and forests; or by intercultural understandings and agreements. To arrive *extra* from across any territorial boundary was a cause for community concern: who is this alien "other," and what is their intent as they come into our accustomed place?

An *alien*, when the term is applied to people, is someone foreign to one's place or to one's experience, whether that experience be cultural or individual. The arriving Other's ethnicity, color, clothing, language, religion, and other physical and cultural expressions are alien to the customs and beliefs of native inhabitants of a place, or alien to the customs of known others with whom they have interacted through trade or regional encounters. Originally (and still presently) *alien* refers to another human being, previously unknown, arriving on the scene. Gradually, as seen earlier, people came to understand that there are nonnative plants and animals in their territory that are alien to their homeland, not solely other humans.

Biblical and American "Alien" Deities

Throughout the world diverse cultures developed the concept of divine beings who could be solicitous for their wellbeing, or demanding of particular modes of conduct from them—or both. Peoples gradually developed specific rituals to honor their deities, principles and laws that were supposed to guide them in obedience to their deities, and specific beliefs about the nature and characteristics of their deities. Formal religious structures or less rigid types of spiritual thinking also emerged. Peoples believed themselves to be in a specific dialogic relationship with the divine beings and other spirits, however understood and described, who became to a greater or lesser extent a significant and guiding factor in their everyday lives and cultures.

It should be evident from this that native peoples subjected to the alien European humans' Discovery Doctrine had their own understandings of the Creator when alien Europeans arrived and sought to impose upon the natives European Christian understandings of God; from natives' perspectives, the Christian God would be viewed as an *alien* deity, much as the Israelites and Jews had viewed deities other than Yahweh as "foreign" or "alien" deities. Natives were not concerned nor did they object to the fact that newcomers had their own understanding of divine being and ritual ways of worship. Natives expected reciprocity in this regard: that newcomers would regard natives' beliefs and practices similarly. They were surprised by and resisted aliens' efforts to convert them to believe in and be subservient to the aliens' deity, a foreign god or an alien culture's understanding of a mutually acknowledged and worshipped Creator God.

When Christian missionaries from the fifteenth century on have tried to convert Indian peoples to their religion—which might have different doctrines in the distinct Protestant churches, and between Protestant churches and the Catholic Church—they were, in effect, rejecting existing Indian spiritual beliefs and practices. Every effort was made to eradicate Indians' "pagan" religion, which was said to include "devil worship," and to force them to worship the "true God." The federal government assisted the Christian religion's enterprise. When Euroamerican-imposed, geographical area-based and geographically limited reservations eventually reduced and even entirely replaced Indians' traditional and extensive lands, Indians were forced to accept missionaries from Christian churches, an imposed alien religion, and conversion of the use of their lands to sites on which churches were constructed as set places of worship and schools were built as locales for (Euroamerican culture-produced) education. Residents on each reservation were assigned just one Christian denomination's representatives to provide religious instruction, so the Indians would think there was a single Christian religion rather than distinct Christian religions with distinct doctrines, in competition with each other for followers. (Indian elders have told me that in times past if you met an Indian and asked what Christian church they belonged to, once they told you, you would know which reservation they came from; and, conversely, if you asked them which reservation they came from, once you had that information you would know the Christian church with which they were affiliated to promote their acculturation.) Native peoples' sacred places in nature, and the sweat lodges and other religious structures they constructed for ritual observances, were desecrated or completely destroyed; Indian elders were murdered outright, forced to become Christians, or even placed in insane asylums. The latter practice occurred on what became the Rosebud Reservation in South Dakota. A Lakota elder showed me where the asylum had been located before it was razed, and spoke about the bones and decayed attire discovered there by accident many decades later: these proved to be remains and remnants of spiritual leaders—"medicine men"—who had been abducted from their communities and never found. In any and all instances spiritual leaders/healers' sacred objects were destroyed, or were "bought" from impoverished relatives coerced into selling them because of adverse circumstances imposed on them—they parted with spiritual objects and attire from revered elders in order to provide for their basic survival needs. Sometimes sacred objects were stolen, including by desecrating Indians' burial places to take them, and placed in museums or put into private collections.

In the Americas, for natives the "true" Creator was the one in whom all people from different Indian traditions believed, whom they worshipped with culture- and area-based rituals and spiritual practices, and whom they regarded and revered in distinct but not competing spiritual ways. Indian peoples' spiritual leaders did not try to convert other Indian peoples to their religion, and neither did they fight religious wars on behalf of their divine Spirit. They respected others' beliefs, recognizing that each had faith in, not absolute knowledge of, the Creator.

In effect, then, for native peoples of the Americas, God, also called Jehovah by some colonists (the divine Being whom some later immigrants called Yahweh or Allah) and their descendants, was a foreign Spirit, an alien deity who was separated into distinct alien deities by those who believed they knew this Spirit's true name and possessed the Spirit's only inspired, sacred book—and each of these sacred books was distinct in different religions, and even in denomination-specific translations. Many Indians, of course, accepted Christianity even when it was not imposed—much as occurred in the time when Jesus' apostles were preaching and teaching in the Roman Empire.

Alien Cosmic Arrivals

The most recent understanding of "alien," highlighted by and perhaps especially catalyzed by the radio version of *War of the Worlds*, is that Earth has had exoEarth visitors: aliens from space. The latter are the most difficult for humankind (and, presumably, for other worlds' intelligent beings where they have evolved and exist) to accept even minimally, let alone to interact with in the long term. Invasive plants can be poisoned or pulled and thereby eradicated; invasive animals can be similarly controlled, whether they actually come from another territory or continent or were the original inhabitants of places to which human populations—and human overpopulation—have expanded; and invasive humans might be restricted or deported by immigration laws promulgated by people who themselves are not native to a place, or by battles waged to protect one's territory from intruders. Aliens from space, by contrast, cannot be so easily resisted or removed, since they have a far more technologically—and presumably, militarily—advanced culture. They might also have pathogens that could decimate human populations, just as alien Europeans had brought smallpox and hepatitis to the Americas, which decimated or entirely wiped out Indian peoples and cultures.

While science fiction novels and films usually portray aliens as Earth arrivals with an acquisitive intent and destructive weapons with which to seize territory and the goods of nature, they rarely present humankind as aliens who arrive on exoEarth worlds with this type of consciousness and conduct. In traditional sci-fi stories, the invading ETI destroy TI cultures and nations; sometimes, a surviving, heroic cadre of humans fights an ongoing guerrilla war against the invaders. The converse is very rarely portrayed: Earth's humans, as TI-become-ETI in space, invade the territory of native TI to seize their lands and the natural goods available on them (which natives had been content to leave in place, use in place, or take from place as needed to provide sustenance), and to subject native TI communities and cultures to human control or destroy them entirely should they oppose the human aliens' imperialistic ventures—as did Europeans when they arrived in the Americas as aliens.

Hollywood has provided a plethora of science fiction films on extraterrestrials over the years. Among them, an engaging young alien in *E.T.* and terrifying invaders in *The War of the Worlds* present contrasting and recurring stereotypes: ETI will be either benevolent (as seen, too, in *Close Encounters of the Third Kind*) or malevolent (as depicted in *Predator*). The majority of sci-fi films portray malevolent, acquisitive, and bellicose *nonhuman* aliens who attack Earth and humans; they do not portray malevolent, acquisitive, and bellicose *human* aliens who attack other worlds and their *nonhuman* intelligent species, and rarely do they portray benevolent nonhumans or humans who "come in peace" to help promote the wellbeing of the resident TI who are less developed socially and technologically. (It seems that the benevolent aliens in either case—humans or others—have a "noninterference policy" when the native residents of a place are harming each other, to let them progress evolutionarily and "work things out"; malevolent aliens, by contrast, are happy to help along the process of species self-destruction in order to further invaders' objectives to acquire territory and goods.)

It seems reasonable to assume on this basis that superior intelligent and benevolent aliens—whether humans in space or nonhumans on Earth—might not attempt to better current conditions on worlds and for civilizations they discover. They might recognize, perhaps through experience, that native inhabitants of a place must gradually evolve on their own to become a socially, economically, and politically integrated community where independently developed diverse cultural expressions are accepted and integrated, and community orientation and solicitude are the norm, honored in principle and practice. This conscious conduct would parallel the ordinary development of biota from birth through adolescence and into adulthood: one does not expect "adult" behavior from a child or adolescent; the younger generations must be educated to that behavior, see it exemplified by others, and incorporate it into their own identity and way of life. To give a loaded pistol to a child or to leave it in a child-accessible place, or to give an adolescent the car keys to drive by themselves without sufficient instruction and supervision, would be examples of actions that would court disaster and endanger for the young person and others.

SETI and NASA

On a more hopeful note, the SETI (Search for Extra Terrestrial Intelligence) Institute in Mountain View, California, founded in 1984 as a part of NASA but now independent, uses radio and optical telescopes to search for signs of ETI, and to prompt an ETI response to human curiosity and the human scientific space quest. *Radio* efforts are linked to the Allen Telescope Array (ATA) at the Hat Creek Radio Observatory, Hat Creek, California, which seeks radio replies from extraterrestrial sources in space. At this writing, the effort is focused on the 2,321 exoplanetary candidates detected by the Kepler Mission using the space-based Kepler telescope. *Optical* SETI research is done in collaboration with scientists from the University of California's Lick Observatory, and with scientists from UC-Berkeley who have linked the Lick Observatory's forty-inch Nickel telescope to a new pulse-detection system that can detect laser beacons sent toward our solar system. SETI also does astrobiology research (primarily seeking simple forms of life beyond Earth) in conjunction with NASA. SETI is associated and collaborates with academic institutions and other research organization units around the globe that use radio and optical telescopes to search for indications that intelligent extraterrestrial life exists elsewhere in the universe. (SETI was popularized in the novel *Contact*, written by Carl Sagan, and in the film of the same name, which was based on the novel and starred Jodi Foster.) On its website, SETI states that "the mission of the SETI Institute is to explore, understand and explain the origin, nature and prevalence of life in the universe. . . . We believe in conducting the most profound search in human history—to know our beginnings and our place among the stars."

Goldilocks Sampled Porridge; Astrophysicists Seek "Porridge" Planets

Astronomical observatories around the world are engaged in complementary research. They look for planets that are in the "Goldilocks Zone"—neither too hot nor too cold in their orbit around their star—which enables them to be habitable, at least in terms of "life" as understood by Earth scientists. In January 2013 scientists at the Harvard-Smithsonian Center for Astrophysics announced that the data they had acquired, from which they had extrapolated possibilities for Goldilocks Zone planets, indicated that seventeen billion planets—that's 17,000,000,000—in the Milky Way galaxy alone are likely to fit their search criteria, and because of that they have at least the conditions for, if not the actual existence of, biota. This does not mean that these planets do have life or that life, if present, has evolved to intelligent life. But the odds are good in the first possibility—that life is present—and the potential is great for the second—that among the billions of planets, intelligent life exists on several.

SETI's predecessor and origin source was the National Aeronautics and Space Administration (NASA), with which it is now an independent co-collaborator. Initially and principally known for its work to put humans in space, on the moon, and on an International Space Station to engage in scientific studies and prepare for voyages to Mars and other planets or planets' satellites, and to launch, maintain, and utilize for cosmic research space-based satellites and long-range optical telescopes such as Hubble, Kepler, and the forthcoming James Webb, NASA has long been involved in a search for cosmic life. NASA's vision is "to reach for new heights and reveal the unknown so that what we do and learn will benefit all humankind."

In 1996, NASA established the Astrobiology Institute to seek and to study origins of life in the universe. "Astrobiology," as described on the institute's website, "is the study of the origin, evolution, distribution, and future of life in the universe. This multidisciplinary field encompasses the search for habitable environments in our Solar System and habitable planets outside our Solar System, the search for evidence of prebiotic chemistry and life on Mars and other bodies in our Solar System, laboratory and field research into the origins and early evolution of life on Earth, and studies of the potential for life to adapt to challenges on Earth and in space." (https://astrobiology.nasa.gov/about-astrobiology/)

In recent years, the NASA Astrobiology Institute (NAI) instituted the NAI Astrobiology and Society Focus Group (NAIASFG), a think tank, of sorts, in which scientists (from NASA and elsewhere), social scientists, and humanities scholars (I am a member of this group) discuss possible impacts on people should NASA or another organization, public or private, discover life, however simple, someplace in space. The group considers and suggests ways by which information about such a discovery should be released to the general public. NAI would have no control over a private company's or a foreign government's announcement or claim that it had scientifically verified proof of exoEarth life; however, NAIASFG might be the vehicle through which the U.S. government would disseminate information that it has gathered on the topic, or it might be the means by which NASA attempts to mitigate negative social consequences of information provided by others and the U.S. government. The multidisciplinary approach of the NAIASFG is intended to provide a broad array of perspectives, including from scientists (geologists, biologists, chemists, astronomers, physicists, etc.), social scientists (sociologists, anthropologists, etc.), and humanities scholars (philosophers, theologians, religionists, historians, ethicists, etc.), to explore collaboratively insights from their respective disciplines that would be informative and helpful to convey to the general public evidence of biota existence. This information, it is hoped, would allay or at least mitigate people's worst fears about the significance and impacts of simple organisms (who might or might not evolve to be complex and intelligent), and of intelligent beings who might provoke, however unintentionally, a human sense of cosmic displacement.

Aliens on and from Earth

Two recent (2009) sci-fi films, *Avatar* and *District 9*, portray exceptions to the usual fictional ideology that we humans, as native TI, are always the "good guys" who must resist the "bad guys," the incoming ETI arriving from outer space to plunder our place. This near-exclusivity of "good guy" status has been presented and perpetuated despite what we've proven ourselves capable of doing to each other when we are all members of the same intelligent species. We are not always "good guys" even to ourselves, even on our own planet. *Avatar* and *District 9* have distinct takes on TI-ETI encounters but offer a similar thematic twist that is contrary and even contradictory to the usual fiction. They suggest that intelligent aliens should be fearful about what humans would do to them both on their distant origin planets and when they come to Earth. Ironically, resident "aliens'" worst nightmares about malevolent human invaders (the true "aliens" on other worlds) are fulfilled in fiction in the two films, much as they were in fact in European humans' Discovery Doctrine, which has devastated native cultures in the Americas since European alien explorers first arrived, and ever since in Euroamerican laws and practices. In contrast, humans' fears about "them," the "Others," the "aliens" arriving from space, have yet to be justified and, hopefully, never will be.

ENCOUNTERING ETI

In Brief: District 9 *and* Avatar

In *District 9*, a massive alien spaceship enters Earth's atmosphere in 1982 (the same year that the historical forced dislocation and relocation of coloureds from District 6 in Cape Town was completed, and the district was subsequently demolished), hovers above Johannesburg, South Africa, and stops, afloat in the atmosphere. Video footage of the ship's arrival reveals that when the spacecraft stopped a small pod had descended rapidly from the craft and hurtled Earthward, burying itself deep underground; despite an intensive search, it is never found. After three months of observing the craft, the South African government sends helicopters filled with troops to enter it; it is so large that the helicopters are minuscule next to it. The soldiers break in, poised to shoot, but find only emaciated and frightened ETI. The extraterrestrial aliens are transported to the Johannesburg region below, where the U.N., in association with the South African government's Department of Alien Affairs, establishes refugee camps similar to those built for human refugees from conflict who flee across national borders. The camp is essentially a shantytown, a refugee camp become concentration camp, with minimum utilities and minimal employment. Initially, the ETI, soon called "prawns" by the South Africans because of their anatomy and because they are regarded disdainfully as "bottom-feeders," are granted deeds to homesteads. Gradually, conflicts develop between the impoverished prawns and South Africans, and the ETI are strictly confined to their camp city. After twenty years, the government decides to remove the prawn population of 1.8 million individuals and place them elsewhere, and contracts with Multi-National United South Africa, a transnational corporation that is the second-largest weapons manufacturer in the world (and has its own private military force), to do the relocation.

MNU Executive Piet Smit appoints Wikus van de Merve, a "not very bright" low-level bureaucrat at the time of the appointment but also Smit's son-in-law (married to his daughter Tania), to oversee the MNU operation. The film then presents a video narrative recorded by Wikus as a historical record of events in and stages of his work. He is joyful going about his task, accompanied by weapons-brandishing soldiers. To make the MNU seizure of prawns' lands "legal," the prawns are "asked" to relinquish their property by signing a mark on a document presented to them. Since most prawns do not know how to read English, when faced with the coercive persuasion of semiautomatic weapons pointed at them they put a mark on the paper.

ETI in Avatar and District 9

Some who refuse are killed; others are said to have "signed" even when they did not do so, on the pretext that some gesture indicated acceptance.

Wikus shares the same racist and condescending attitude toward the prawns held by other South Africans. He oversees the paper-signing process, in the company of MNU soldiers and his assistant, Fundiswa Mhlanga, a black African. Although Wikus and the soldiers wear body armor, Fundiswa is not given any. Wikus assures him that he'll be okay. When a nest of prawn eggs is burned, Wikus gleefully describes the "popcorn popping" sound emitted.

Wikus, the MNU civilian staff member who is supposedly heading the operation, soon learns that the MNU commanding officer, the battle-seasoned, well-armed, and cruel Colonel Koobus Venter, is really in charge. MNU is interested not only in receiving payment for its services but also in finding caches of prawn weapons. Disclosure of this plan and endeavor reveals MNU's real focus and goal in the relocation campaign. They already had found some seemingly powerful weapons, but they appear to be linked in some biological fashion to the prawns and can only be operated by them. The mercenaries have a policy of "shoot first and ask questions later," according to Venter; this attitude and action would be strengthened by prawn weapons. MNU is secretly experimenting on kidnapped prawns in labs, sometimes while they are still alive, to try to develop a hybrid mix of their DNA with humans' DNA, or to transplant selected prawn body parts to humans, in order that selected humans—MNU soldiers—could fire ETI weapons.

In the course of carrying out his duties, Wikus accidentally sprays himself with fluid from an unmarked can as he is ransacking the home of Christopher Johnson, an unknown prawn scientist and leader. Johnson has a primitive yet technologically sophisticated laboratory beneath his home. There he has been manufacturing parts and repairing equipment and parts found in a trash heap to restore the spaceship and make its defunct computers operable. Under his shanty, too, the escape pod from the spaceship is buried and hidden. From partially filled cans they have found, Johnson, his precocious and highly intelligent son, and a friend have secretly and gradually been harvesting fuel to run the ship.

Wikus, to his horror, soon discovers that his arm and hand are being morphed into a prawn appendage. As the film progresses, he becomes more and more an ETI, and very sympathetic toward them; meanwhile, MNU and the South African government declare him a wanted man after trumping up charges against him. MNU wants to dismember Wikus alive to transplant his fresh body parts to their own men, who would then be able, they assume, to fire ETI weapons. Wikus discovers that the long lost pod from the spaceship is buried under Johnson's shack, and when Johnson says that he would be able to restore Wikus to his human body, Wikus agrees to help him escape with his son to the spacecraft.

The remainder of the story, filled with twists and turns and additional character development (including of a psychopathic Nigerian warlord), can be condensed to noting that as Johnson and his son prepare the pod for flight, MNU mercenaries try to stop them; Wikus provides firepower to cover their escape, using against the MNU soldiers a variety of ETI weapons: his partial transformation into an ETI gives him sufficient DNA to operate them. Father and son arrive on the spaceship, activate it, and take off. At film's end, a sorrowful Wikus, completely ETI now, stands on a hill overlooking Johannesburg and laments his personal losses: his wife and his people. Now a much-wanted man, he lives among the ETI, accepted by them as one of their own, but hoping for Johnson's return. Fundiswa is imprisoned for being a whistle-blower because he goes public about MNU's illegal operations. (I encourage the reader to see *District 9* in its entirety and, besides enjoying the action-packed story, note the underlying themes guiding it.)

In *Avatar*, Jake Sully, an ex-Marine corporal whose wounds made him a wheelchair-confined paraplegic, is asked to replace his recently deceased twin brother, a research scientist studying the Na'vi, a local branch of the Omaticaya people, on Pandora, a moon in the Alpha Centauri system. Jake is attracted by a promise of substantial financial remuneration and, after he arrives, a pledge that the RDA Corporation will provide him with the resources to have surgery that would enable him to walk once again—a surgery that veteran's benefits would not cover. He is chosen to join the scientific research team studying the Na'vi—as an avatar whose Na'vi body is infused with his brother's DNA. The head of scientific research, astrobiologist Dr. Grace Augustine, is unhappy with this choice, although the DNA allowed no other possibility: she wanted a scientist, not a soldier. Once situated, Sully is told by the RDA security officer and mercenary leader, Col. Miles Quaritch, that he will secretly report to Quaritch to relay information about Na'vi defenses and living space.

Beneath Na'vi land lies a vast deposit of a rare mineral, unobtanium, worth $20 million per kilo; RDA wants to strip-mine it. If negotiations compelling the Na'vi to move are not successful, Quaritch will lead an assault on their community to kill or disperse them. At first, Sully does as he is instructed. But then Neytiri, a Na'vi woman, saves his life when he is surrounded by a fierce pack of wolf-like animals. Previous to this encounter she had seen him from a perch high in the trees, and was about to kill him with her bow and arrow when a jellyfish-like creature alights on the arrow; she sees this as a sign to spare him. At the Na'vi village, Sully is greeted first with anger, and Neytiri's father, the Na'vi chief Eytukan, is about to order Sully's execution but his wife Mo'at, the spiritual leader of the Na'vi, intervenes. She says that he has a special role to play for the Na'vi, and instructs Neytiri to instruct Jake in Na'vi ways.

Jake goes back and forth between the RDA base and the Na'vi village: by day he is an avatar in the jungle, able to run, climb, and jump; at night he is at the base in his wheelchair, reporting back to Quaritch. In the jungle, at first Neytiri calls him a "baby" and a "moron" because of his ignorance of the ways of the forest, but gradually, as he learns the Na'vi language and cultural mores, she accepts him. He becomes a Na'vi warrior, rides Pandora "horses" and flies on gigantic birds. Their love grows and they pledge themselves to each other beneath the sacred Eywa tree. Sully's former boss and commander see him protecting the Na'vi community when he tries to stop a gigantic bulldozer from continuing to topple the tall forest trees. In the first major battle, the Na'vi arrows, their aerial assaults flying on their birds, and their ground assaults riding on their "horses" are of no avail against the well-armed RDA military in helicopter gunships that launch missiles and fire machine guns, and RDA mercenaries in full battle dress, some riding in giant robots equipped with multiple weapons. A missile attack launched by a large aerial battleship cuts the roots of the massive Hometree, and it topples. Many Na'vi are killed in the battle, and many more are wounded and culturally and spiritually dazed. Grace Augustine dies from wounds received. The Na'vi learn of Jake's deception and betrayal, and he is cast out. He realizes that he must prove himself, and unites with the fierce Toruk, a giant predator bird called the "Last Shadow" because once someone sees its shadow, they'll soon be eaten. He returns to the Na'vi on Toruk, and immediately they accept and respect him; he becomes their leader. Messages go out to all the Omaticaya clans, and the Na'vi forces are strengthened substantially when other clans' warriors arrive. Jake goes alone to Eywa, the Tree of Souls where the spirits of all who have died now live; it has a biological connection to all life on Pandora. He prays hesitantly to Eywa, asking for help. Neytiri overhears him, and tells him that Eywa does not choose sides, but strives to keep balance on Pandora. The second great battle is engaged, and things are not going well for the united Omaticaya people. When defeat seems inevitable and imminent, suddenly the ground trembles as all the fiercest of the Pandora fauna charge out of the forest against RDA forces. Col. Quaritch is killed by arrows from Neytiri's bow as he is about to kill Jake. The Omaticaya and their new allies send the alien ETI humans back to Earth. Jake and two of the scientists remain behind to live new lives with the Na'vi on Pandora.

Parallel Themes, Schemes, and Scenes

Avatar and *District 9*, released almost simultaneously, are linked through New Zealander Peter Jackson, who was the director of *Avatar* and a coproducer of *District 9*. Although lacking apparent mutual influence, they have several parallels.

In *Avatar* humans, the invading ETI, intend to plunder planet Pandora, the territory of the native TI, the Na'vi Clan of the Omaticaya people, and displace them to acquire a desired mineral, unobtanium. The RDA conglomerate wants to mine unobtanium even though it lies beneath the Na'vi village located in and around their Hometree, and beneath the nearby site of their sacred Tree of Souls. *Avatar* has obvious parallels to the history of European colonization of the Americas, during which indigenous peoples, North and South, were subjugated by the sword to the cross and the crown. In *District 9*, humans are the native TI but the incoming ETI are not intent on plundering Earth. Rather, they are essentially malnourished immigrants who need to be rescued from their massive, disabled spacecraft. Because of their insectoid appearance they are disparagingly called "prawns." Gradually, to some extent, they become TI, and adapt to Earth attitudes and customs. Essentially imprisoned in a refugee camp-become-concentration camp they are not assimilated, but most are losing to some extent their own culture and cultural identity. Many if not all yearn to go home, to be ETI on Earth once again and TI on their own, distant world. They, like native Africans in past centuries, are being displaced so that South African whites and the Multi-National United transnational corporation will benefit financially—at their expense.

In *Avatar* the co-protagonist, Jake Sully, is a mix of human and Na'vi DNA as an avatar on Pandora, and gradually comes to sympathize with and support the Na'vi against his former mercenary comrades. In *District 9* the protagonist, Wikus, breathes a spray that gradually transforms him into a mix of human and prawn DNA. He begins to transform from a racist white South African into a mixed-blood hybrid of both human and prawn species, becomes sympathetic to and supportive of the ETI prawns, and fights with them against his former employer and the mercenaries they use as private military contractors to oppress and kill or relocate the prawns.

Racism or at least ethnocentrism is evident, too, in the ways humans interact with and treat ETI. The corporations and mercenaries in both films are disdainful of the "other." In *Avatar* Parker Selfridge, the corporate executive leading the mining expedition, calls the Na'vi "blue monkeys" and "savages," and declares, "Look, killing the indigenous looks bad, but there's one thing that shareholders hate more than bad press, and that's a bad quarterly statement." He calls them "flying savages who live in a tree" and states that if their sacred tree, Eywa, is destroyed, "They can relocate." When he is told that Eywa and the land around it is "sacred," Selfridge says dismissively, "If you throw a stick in the air around here, it's going to land on some sacred fern." Colonel Miles Quaritch, head of security and the mercenaries' commander, calls the natives an "aboriginal horde" and "savages"; their home is "the hostiles' camp." ("Savages" and "hostiles" were derogatory terms used by Euroamericans in the nineteenth century to refer to Indians who were living, farming, and hunting in their treaty-reserved home territories and resisted theft of their homeland by Euroamericans when U.S. soldiers and settlers occupied native lands.) When the Na'vi flee from helicopter gunships firing missiles at Eywa, he says, "That's how you scatter roaches."

In *District 9* the word *prawns*, said to connote "bottom feeders," is used disparagingly against the arrived ETI. Just as occurred when Europeans invaded native territories around the world, the South Africans have not learned the ETI's name for themselves; they are always the prawns. Throughout the film, this continues to be the case: the "otherness" of the "other" is constantly reinforced through repetitive use of prawn. The practice of "legalizing" land transfer through duress placed on an ethnic minority to sign an agreement that enables a dominant minority to take their property did not originate or occur only in apartheid South Africa; historically, as with other events in the film it has been done elsewhere, and globally. The document-signing process parallels events in the U.S., for example, when Indian peoples were forced by U.S. military-imposed treaties to relinquish vast territories to the federal government. The natives, accustomed to oral traditions, did not know how to read; they had to trust in what was read to them: that they would retain rights to and limited sovereignty over a small segment of their original territories, or that if they gave up their lands and accepted relocation, they would be provided with a comparable, habitable place elsewhere, sometimes almost two thousand miles from their homeland. Such "treaties" were the "legal" justification—based on the Discovery Doctrine—for U.S. government actions. The Trail of Tears across the southern U.S., from the southeast to the southwest, was one example of forced relocation. The reduction of Lakota lands in South Dakota from more than half of the state to two small reservations in the southwestern corner, and of Northwest Indian nations' lands in Washington State from ten million acres to one million, are examples of Indians' confinement to a minimal part of their original territories.

In both films, there are analogues to the dangerous "military-industrial complex" against which President Dwight D. Eisenhower warned U.S. citizens: as Multi-National United (MNU), a transnational corporation, in *District 9*; as Resources Development Administration (RDA), a transgalactic corporation, in *Avatar*. The operative corporate exploitative capitalist ideology is dominant on both worlds. In *Avatar* it underlies both Earth's demise and the threat of a like devastation of Pandora if the Na'vi do not successfully resist corporate extractive policies. As Jake said to the assembled Na'vi, "The Sky People have sent us a message: They can take whatever they want." In *District 9*, the prawns' territory and homes are ruthlessly obliterated by the MNU military operation.

Ecological destruction is presented in both films, more graphically in *Avatar* where humans burn the forest and kill biota—plants, animals, and intelligent beings. Jake Sully comments at one point that humans, the "Sky People" in Na'vi nomenclature, "killed their mother," Earth; after the Omaticaya unite from around Pandora to defeat the human mercenaries, Jake says of the humans that "the aliens went back to their dying world."

A profound character cultural change is presented in both movies. In *Avatar*, Jake Sully transforms from an ex-marine transgalactic corporation's mercenary to a warrior leader who has learned Omaticaya traditions—including spiritual understandings and respectful relations with Pandora's biota (such as by apologizing—which North American Indians did—to the animal he killed to provide food for the Na'vi), and appreciation for the integrated Pandora biotic network—and fights on behalf of the Na'vi. He comes to realize that his external and internal conversion, and his consequent participation with the Na'vi in their battle against his native species, former comrades, and former employer, gives him a new, twofold status as an *alien*: he is simultaneously an alien to his original species—he calls himself "outcast, betrayer, alien"—and an alien to the Na'vi. In *District 9*, as Wikus morphs into a prawn, he experiences physiological, psychological, and philosophical transformation. His bodily changes are obvious, but he also comes to understand and be sympathetic to the prawns once he sees that among them, just as among South Africans, there are highly intelligent individuals, friendship bonds, and familial relationships. Wikus becomes TI-as-ETI, in head and heart, and eventually he is ETI—an alien—on his home planet, physically and socially, while longing to be TI on Earth once again.

Avatar makes obvious allusions to the bombing of the World Trade Center on September 11, 2001, and the U.S. bombing of Baghdad in its aftermath (even though no Iraqis were among the nineteen terrorists who took over civilian aircraft to destroy the WTC, the Pentagon, and the White House: fifteen were Saudis, as was al-Qaeda leader Osama bin Laden, who ordered the attack). In *Avatar* Colonel Quaritch states when the Na'vi clans unite to fight rather than capitulate to RDA demands, he and his troops will launch "a preemptive attack. We will fight terror with terror." An avatar program scientist calls Jake and warns him that "some kind of shock and awe" attack was being prepared, the very phrase that President George W. Bush used when threatening to bomb Baghdad if Saddam Hussein did not surrender. Colonel Quaritch's subsequent assault on the Hometree to destroy it, demoralize the Na'vi, and remove an obstacle to mining unobtanium, recalls the WTC attack. The *Avatar* message in both instances might be interpreted via a phrase in the old newspaper cartoon *Pogo*: "We have met the enemy, and they is us." New York residents in particular might see themselves in the scene where Hometree is destroyed—by U.S. mercenaries—and realize that the U.S. does not have "clean hands" when it verbally and militarily attacks "terrorism" at home and abroad. Similarly, U.S. citizens as a whole might realize that U.S. "shock and awe" bombing in real life, as seen live on television or later in televised news reports, harms civilians and civilian homes: not all are a real or imagined "enemy," a dangerous "other." While "shock and awe" might sound great when you're its perpetrator, it sure is hell when you're its recipient, whether Iraqi or Na'vi, as fiery scenes from nonfiction Baghdad and fiction Hometree amply demonstrated.

Aliens in Avatar: Human and Native

When *Avatar* was released in theatres around the globe, public interest in extraterrestrial exploration and its consequences was stimulated once again. This time the stimulus was provided by futuristic fiction about a distant world, not by statements regarding present-era factual Earth events such as presented by witnesses' descriptions of Roswell crashes. *Avatar*'s advanced special effects enhanced its story, and in a dramatic way invited people to consider deeper implications of TI-ETI interaction. Right-wing ideologues writing in newspapers such as the *Wall Street Journal* berated the movie as an attack against capitalism and "free enterprise" because of its portrayal of the military-industrial complex voraciously in action on a distant world, and its underlying implication that similar practices were in place already on Earth in the present. Some disliked its suggestion that in space military-industrial complex representatives would do to the native TI on Pandora what colonists did (and their descendants are doing) to indigenous peoples on Earth.

Beyond promoting appreciation for the latest technological enhancements available to tell a story, *Avatar* has the potential to provoke serious reflection on real-life implications of terrestrial-extraterrestrial Contact—and terrestrial-terrestrial contact as well. *Avatar*'s 3-D technology and computer-manipulated actors and scenes are fascinating; its warnings about Earth-based explorers' and entrepreneurs' potentially catastrophic impacts on other planets' environment, biota, and indigenous cultures are thought-provoking. One of the topics explored minimally and superficially in the story is potential science-religion conflict about or complementary concern for even the idea to establish Contact with ETI, let alone using available technology to do so. In *Avatar* Na'vi spiritual beliefs are discussed derogatively by Selfridge and Quaritch. Even Dr. Grace Augustine, a leading scientist who heads the scientific project and is sympathetic to (and empathetic with) the Na'vi, while commenting positively on Na'vi spirituality seeks to explain it solely in scientific terms. She provides a scientific rationale and data for their beliefs and practices; at several points in the film she notes her lack of belief in spiritual concepts, which contrast with her beliefs in science constructs.

Dances with Wolves and Avatar

Avatar's storyline has an obvious link with the movie *Dances with Wolves*. Both films portray impacts on native intelligent people when aliens—in the form of nonnatives with a contrasting culture, powerful military, and imperialist ambitions—invade their territory. In *Dances with Wolves*, in the nineteenth century Lieutenant John Dunbar, a former U.S. cavalry officer, travels west and comes to befriend his new neighbors, Lakota Indians who have long resided on the land. He gradually becomes assimilated with them, primarily through his developing relationship with a white woman, Stands With a Fist, who was adopted by the Lakota as a child decades earlier. She is the daughter of Kicking Horse, Lakota spiritual leader and warrior. In *Avatar*, Jake Sully, a paraplegic marine who can walk as an avatar (a mix of human and Na'vi DNA) on Pandora, is befriended by Neytiri, his co-protagonist in the story, whose father is the Na'vi community leader Eytukan and whose mother is the Na'vi spiritual leader Mo'at. In *Dances with Wolves*, Dunbar becomes sympathetic to the plight of the Lakota, who are suffering loss of their land and goods because of an ongoing Euroamerican invasion supported by the U.S. cavalry. He partners with the Lakota against the U.S. cavalry, with which he had previously served. In *Avatar*, in a century far in the future, co-protagonist Jake Sully, a former Marine, in his avatar form on Pandora comes to oppose his former employers and paramilitary unit and to fight with the indigenous Na'vi. In other parallels, Indian actor Floyd "Red Crow" Westerman's character in *Dances with Wolves*, Chief Ten Bears, is reprised in Indian actor Wes Studi's portrayal of Eytukan, the father of Neytiri; the initially antagonistic warrior Tsu'tey in *Avatar*, who eventually becomes a friend of Jake, re-presents the initially hostile Wind in His Hair (portrayed by Wes Studi), who became Dunbar's friend by the end of *Dances with Wolves*. (People who liked *Avatar* would do well to watch *Dances with Wolves*, and vice versa, if they have not seen both; doing so would stimulate attention to a past-future historical complementarity, and in-depth knowledge of the films' related social insights.)

Stephen Hawking and Avatar

Stephen Hawking stated that humans should soon have a moon base and a Mars colony to avoid human self- and planetary destruction on Earth. Humankind, too, he said, should shut down SETI and hide from ETI on

Earth and on distant worlds because ETI likely have destroyed their home planet and seek new places to conquer and exploit. His assessment and remarks contrast sharply with *Avatar* in that he expects humankind not to replicate in the stars what they've done on Earth (which humans do on Pandora), and are congruent with the *Avatar* narrative in the assertion that ETI (humans on Pandora) likely would have destroyed their origin planet and are intent to do likewise to others' worlds elsewhere in space.

In his speculation, Hawking is pessimistic about the consequences of evolved intelligence in itself: humans are destroying their Earth home and should desert the neighborhood; aliens have destroyed their home planet and are relocating elsewhere in space. While effectively Hawking declares that ETI will perpetrate a Discovery Doctrine in space, he does not consider that humans, too, might be such aliens—as European humans were when they arrived in the Americas—even though he had noted also that the attitudes and actions of Columbus and his successors "didn't turn out very well for the Native Americans."

In contrast with Hawking's positive portrayal of what humans would do in space, *Avatar* is more historically informed and realistic when it projects what will be probable human conduct on distant worlds. It recognizes that historically on Earth human ingenuity often has employed terrestrial technology with few ethical constraints, thereby increasing Earth's devastation; acquisitive and technologically superior humans in space will act throughout the cosmos as they have acted on Earth.

Hawking's *deus ex machina*—spaceships and artificial habitat colonies that will rescue some people from Earth, a planet they have devastated and are increasingly destroying—will enable humankind to do better in the heavens than it has done on Earth. He does not suggest the need for, let alone development of, a more responsible human consciousness, conscience, and conduct on Earth in order to ensure that humankind does better in the heavens than it has on Earth. *Avatar* shows what will likely happen should humankind venture forth ethically challenged and unchanged. The consciousness, perspective, and ethics that humans take with them on interworld expeditions will guide their treatment of extraterrestrial biotic and abiotic being. *Avatar* illustrates this dramatically. It shows what types of social and ecological contamination and catastrophes will result if unaltered human consciousness is expressed extraterrestrially as it is terrestrially. As Sully comments in *Avatar*, the Sky People (humankind on Pandora) "think they can take whatever they want." *Avatar* teaches that

Earth and the extended cosmic creation do not need an idealized human dis-placement to idealized new worlds, but rather real humans' *a priori* ethical transformation for all present and future places.

Hawking expresses as a scientist what *Avatar* portrays fictionally and what European expansion into the Americas illustrated historically—ecological, political, and social devastation. He notes that "we only have to look at ourselves to see how intelligent life might develop into something we wouldn't want to meet." Hawking doesn't continue his admonition that we should "look to ourselves" and notice what humankind has wrought on Earth, and what humans likely would do in space, on other worlds where TI natives live, when ETI humans arrive.

Avatar's warnings, and its promotion of better human conduct, provide a corrective for Hawking's naiveté. Reflection on both *Avatar*'s scenes and Hawking's scenarios might prompt people to act responsibly on Earth and in the cosmos beyond, and stimulate a mutually beneficial dialogic relationship between present and future time, and among distinct but linked contexts and cultures.

Aliens in *District 9*

Scientists have predicted that in the event of a nuclear war or some forms of natural catastrophe on Earth, only insects would survive, principally cockroaches. In *District 9*, ETI who have come to Earth are insectoid. Perhaps there's a subtext suggestion that on their own planet the prawns had a major war or natural disaster. The only biota that survived were insects. They evolved over time to have an anatomy and physical characteristics that enabled them to manipulate tools, which stimulated in turn emergence of a higher intelligence and the development of science and technology.

Many viewers of *District 9* find the alien prawn species repulsive at first. Their worst characteristics and conduct, as they adapt to their confining environment, are portrayed: through the lens of the camera and the pens and commentaries of members of the controlled South African news media, and in descriptions of them by representatives of the military-industrial complex, Multi-National United (MNU), which is forcibly relocating them using private military contractors. However, most viewers come to see another perspective on the aliens as the movie progresses. They, too, have scientists and technologists in their community, are compassionate, and have family bonds—as is illustrated especially in the relationship between a scientist father and his precocious young son. (Today, parents in cultures permeated and seemingly dominated by computers can well appreciate the young prawn's computer literacy, skills, and expertise. Some adults might see in him their own or neighbors' children who are similarly capable—while the adults can barely make a VCR or DVD work even to play a movie like *District 9* or *Avatar*, let alone make a computer execute complex sequences.)

District 9 is an unusual movie in that its narrative relates events that occurred in a *past* period, from 1982 to 2002, rather than a *future* one, as is usually the case with sci-fi films. After all, people who were around in 1982 and are alive today would not be able to recall that such a dramatic event, involving a space vehicle, physically distinct aliens, and major conflicts between TI and ETI had occurred and ended just seven years before the film's release. The reason for this chronological disconnect is presented subtext in the title, historical period, and social intent of the movie. Like *Avatar*, the movie provides not only a fascinating sci-fi film, but also a social commentary on current events, in particular the treatment of minorities regarded as "Others"; its special focus is apartheid South Africa (thus the anachronistic date). In *Avatar* the oppression of and injustice toward indigenous peoples that characterized the *past* Discovery Era and continues into the *present*, is portrayed continuing into the *future* if human consciousness and conduct remain unchanged and the military-industrial complex is given license to exercise unrestricted political, economic, and military power—and control nations, behind the scenes. In *District 9*, in contrast and yet complementarily, the focus is on injustice toward ethnic groups in the recent *past* and *present* who, whether or not they constitute the majority population, are a political and economic minority, and therefore relatively powerless against the well-armed, politically powerful, and economically dominant minority population. In *Avatar*, future humans devastate a distant world, Pandora, and try to commit genocide against a physically distinct ethnic group, the Na'vi, Pandora's sole intelligent species. In *District 9*, contemporary humans destroy a community of aliens in an area of Earth, and reinforce the apartheid policy that harshly oppresses a terrestrial ethnic group, the "coloureds," who are members of the same human species, with a distinct color and different ethnic culture. (White South Africans, of course, have a "distinct color" relative to majority black South Africans.)

Avatar, *District 9*, and Discovery Doctrine

Both *Avatar* and *District 9* illustrate the use of Discovery Doctrine against extraterrestrial intelligent species. A principal element of Discovery was that Europeans had a "superior" civilization when compared to any other people. A superior civilization was identified in part by its development of a private property economic system, where land was marked by boundaries and worked individually by farmers in small family holdings, or by renters on vast estates owned by the monarch or nobles. This economic arrangement satisfied the wealthy minorities' greed for an ever greater accumulation of wealth, at the expense of the common people who sought sustenance and security. Many peoples throughout the globe had a community-oriented, communal society with communal holdings in land; this cultural practice automatically failed the European "civilized" requirement.

With Discovery Doctrine in mind, note how events in both *Avatar* and *District 9* illustrate Discovery in action. Both films warn against—without citing it directly—continuation of the Discovery ideology underlying historical events that initially involved European governments, then Euroamerican governments and judicial systems, and later other governments around the world, and became incorporated into international jurisprudence. The films suggest that arrogant and aggressive humans and those under their command will oppress in the future, as they have in the past and present, less technologically developed and/or politically powerful entities and ethnic cultures—whether these be found on Earth or in the heavens.

In *Avatar*, human invaders organized and equipped by Resources Development Administration try to impose their "superior" culture and civilization on the supposedly backward and culturally "inferior" Na'vi people. The human aliens, the invading ETI entering the territory of the native TI Na'vi, try to replace the traditional, communal Na'vi communities, religion, and spirituality with a new religion of capitalism and its individualistic ideology based on competition and greed. (Harvard theologian Harvey Cox describes the capitalist religion—its beliefs, rituals, clergy, and doctrines—insightfully and thoroughly in his March 1999 *Atlantic* article, "The Market as God—Living in the New Dispensation." He shows how the Market belief system, with its Bear and Bull idols, has more adherents than any traditional theist faith and is an unrecognized threat to all of them.)

In *District 9*, the South African government and its dominant partner, Multi-National United, use paramilitary mercenaries to displace the prawns, who are supposedly culturally inferior (despite their technologically superior weapons and massive, technologically superior spaceship). The ETI's Earth territory, on which they had lived for twenty years, is to be taken from them; they would lose their homes. Whatever limited livelihood they had developed on site would be severely disrupted, if not completely eradicated, at displacement. The ETI have no apparent religion, nor do the TI in the film; however, the TI capitalist creed and greed, coupled with military advantage, are evident in *District 9*, just as they are in *Avatar*. MNU's main intent is not simply to help the South African government's forced relocation of resident ETI, but to seize ETI weapons and be able to use them to defeat any other military forces and dominate nations when paid to do so. Since the weapons are far superior to any human-developed arms, they would be very useful commercial products whose sales would profit MNU at home and abroad. MNU is Earth's second-largest arms manufacturer and supplier. Its mercenaries would use ETI weapons when they contract for service on behalf of foreign nations; MNU would sell arms to other nations and mercenary groups not in competition or conflict with MNU contractors.

Apartheid and District 9

District 9's link to apartheid is very direct. The storyline of the film was inspired by an apartheid-era relocation of some sixty thousand "coloureds" (people of mixed race) in District 6 (note the inverted 9), near Cape Town. The South African government claimed that District 6 was a hotbed of illegal activities: crime was rampant and drugs and prostitution were readily available. (Critics of the apartheid government's actions said that the real reason was to convert District 6 into a tourist region.) From 1968 to 1982 the forced relocation emptied the District, and all buildings were razed except the Christian church. Once apartheid was eliminated, and the majority population had control of the country under Nelson Mandela and the African National Congress (ANC), the situation changed dramatically. In 2002, on the twentieth anniversary of District 6's destruction, the ANC government declared that the people who had been forced from their homes and places of livelihood in District 6 had a right to return and reclaim them.

The story of the ETI immigrants denigrated as "prawns," then, recalls apartheid injustices suffered by the earlier inhabitants of District 6, who had been despised in white South African eyes because they were "coloured," an inferior-to-whites racial/ethnic mix. Reflection on District 6 events reminds us of how we have treated the "others" of our own species. We are advised to recall that conduct and are admonished not to similarly mistreat extraterrestrial intelligent species if they call "home" a territory that humans desire, in itself or because of the natural goods available in it. *Avatar* tells a similar tale and prods us, as does *District 9*, to rethink what we have been doing on and to Earth and even to Earthlings, who are physically akin, and genetically kin, to members of the dominant ethnic group that does not responsibly relate to members of its "own kind" and likely will act in a similar manner toward "otherkind" in space.

Avatar and *District 9*: Eco-racism?

Sociologist Robert Bullard has analyzed occurrences of racist environmental abuses in the U.S. and presented his data in several books. In *Dumping in Dixie: Race, Class, and Environmental Quality*, Bullard studies particularly the siting of toxic waste burial and toxic pollution-emitting industrial plants near minority (especially African American) residences—of all social classes, not solely the poor. He discusses, too, the siting and construction of suburban middle- and upper-class residences on places that previous users and developers knew had been polluted with toxins; the homes were offered to African American families. As family members become ill from toxic residues, they are forced to move, and try—with little or no success—to be reimbursed or recompensed for the expensive homes they have had to leave, and for their medical bills. Most people on these sites had used their homes as their principal investment and the principal depository of their life savings.

A similar situation has arisen in urban areas. Waste-burning companies have attempted to site their toxin-producing facilities primarily or solely in predominantly ethnic or racial minorities' places such as Harlem in New York City, and East Los Angeles. Their callous efforts catalyzed coalitions of local residents, local churches, and environmental organizations to fight against the siting and constructing of toxic waste incinerators in their neighborhood. The public's interest to protect their person, property, and social wellbeing prevailed against private interests promoting excessive profits. Then there is Cancer Alley, between Baton Rouge and New Orleans, and other places where efforts to stop ecological and economic harm are ongoing, and people's resistance to being poisoned, by toxic emissions and effluents, although begun, has not stopped all harm.

If some sort of private interest in profit over public wellbeing were to be attempted in space, it might be motivated not only by financial greed but, however unacknowledged, by racism. *Avatar* portrays this openly when Parker Selfridge and Colonel Quaritch call the Na'vi "blue monkeys"; in *District 9*, the ETI who arrive in South Africa are called "prawns." There is obvious eco-racism in *Avatar* when, in order to extract subsurface unobtanium, RDA brutally destroys the Pandora environment, Na'vi homes, and sacred sites in the areas around where the Na'vi live. In *District 9*, eco-racism is more subtle: the ETI are told that they will be moved to a better District. Wikus tells Christopher that this is a lie. He adds that the houses have a more cramped living space, and the new District is like a concentration camp. After Christopher and his son leave in the ETI spaceship, District 9's biotic, cultural, and social environments are destroyed. The racist economic intent of the South African government and the racist profit desires of Multi-National United are satisfied. The disposition of powerful ETI weapons, however, is unresolved at film's end.

Continuation of eco-racism "in the heavens as on Earth" is a real possibility if people on Earth do not eliminate their eco-racist consciousness and conduct here on our own planet. Members of the public who oppose such an ideology and its practices must remain diligent—as they were in opposing eco-racism on Earth—in analyzing space exploration, colonization, and industrialization proposals and projects.

If human minds and hearts, and attitudes and practices do not change, negative consequences will result on Earth and in space when TI-ETI Contact occurs. A noticeable anatomical, cultural, or other obvious distinction between humans and ETI might be the rationale for some humans to continue "business as usual." This would eliminate prospects for TI-ETI cooperation and collaboration—to human detriment. As noted earlier, ETI civilization might well be billions of years older than TI civilization, and possibly could assist us in several beneficial ways (medical and technological, in particular). It is in humans' self-interest (domestically and cosmically) to evolve consciously beyond evolutionarily primitive ways of thinking, such as are evident in Earth racism and humans' destruction of their own planetary home. Beyond self-interest, it is "the right thing to do."

Recognition of past histories of xenophobia, political imperialism, and even genocide or attempted genocide against an "Other," the *Avatar* and

District 9 filmmakers suggest, might prompt us to realize that we humans must change our dominant consciousness and conduct on Earth before we venture into space; otherwise, we will act toward other intelligent species as we have acted toward select groups within our own intelligent species. Continuing our harmful practices on Earth will result in wars among ourselves that will increase in number and ferocity. We will continue to lack the cohesion to be a community of explorers who are involved in collaborative projects to explore the cosmos and to extend our places of habitation to meet the needs of our expanding population. We might also, by continuing "business as usual," alienate other intelligent beings—which will have adverse consequences when they prove to be the "superior" civilization, perhaps one in which a communal way of life and community orientation are fundamental characteristics. They might even have a superior form of "spirituality" that unifies them as a species, instead of competing religious institutions whose beliefs divide them in the way that the human species' institutional religions and religious doctrines divide us.

Future Possibilities: ETI and TI in Cosmic Community

If ETI flybys described by credible witnesses indicate that intelligent aliens are studying us scientifically and seeking to gauge our level of species social maturity and ecological responsibility, our current conduct and the type of consciousness it reveals would caution them about making Contact and initiating a long term relationship at our present historical moment and evolutionary stage. In some ways, our evolution has stagnated. We have not been able, as one human family and a community of communities, to recognize that we are interrelated, interdependent, and potentially collaboratively integrated. Until we arrive at that new awareness (or old awareness become societally operative once again) and act in a corresponding manner, we will be unable to "get our act together" in order to explore, as TI and with ETI, the cosmos that is the broader setting of our common existence, and to share its wonders and goods for our mutual benefit and wellbeing.

Whether or not ETI is observing us, and whether or not we have sustained Contact with ETI, we need, for our species and planetary survival, a new cosmic consciousness, a cosmic concern for the wellbeing of the cosmic community, and a consequent commitment to concrete cosmic caretaking—beginning with our own world and its inhabitants. We should not—cannot—export into space our destructive social and ecological practices. *Avatar* and *District 9* have warned us that we will do so if we do not have a socioecological conversion that results in our having a relational community consciousness—on Earth and in the stars. Before we colonize elsewhere, humans should weave together technological expertise, socioecological (social justice and environmental wellbeing) ethics, and a cosmic concern for the social, cultural, and physical wellbeing of the biotic community in its diverse forms—and resolve to avoid harm to extraterrestrial locales and life. A dialogic relationship could be established then between present Earth and future Earth, and between Earth and other worlds, present and future. Reflection on and repentance for what humans have done to Earth and, in the case of European colonization of the Americas and Africa, to Earth's indigenous populations and ecosystems, might lead to an ecological (if not spiritual) conversion.

People who have seen *Avatar*—and those who have not—if they reflect on and repent for human harm to Earth, oppression of indigenous peoples, and extinction of multiple species, will reject the disrespectful, disruptive, and denigrating attitudes and actions humans express in the movie. Humankind will strive, if that happens, to accept the Other discovered on other worlds and the Other encountered at home on Earth, and to treat them as coevolved members of a single and singular intelligent life community. Should that happen, ethnic and species distinctions will disappear, replaced by acceptance of each other, because we have acknowledged that we are all evolved stardust become complex and self-reflective, and thereby potentially able to adapt to and be integrated within both the Earth community in which people are already related and the cosmos community in which all intelligent beings are related.

An active and harmonious ETI-TI interrelationship will come to pass when humankind and otherkind, all intelligent beings, are committed to collaborative, relational communities. Native populations on other worlds will not in the future, then, be subjugated and oppressed by humankind as indigenous populations have been on Earth, and natural ecologies on other worlds will be conserved, not contaminated or excessively consumed. RDA should not be a business model for Earth-based corporations exploring space. Humankind, as it considers the extreme ecological destruction it has wrought on Earth and the social devastation it has caused for diverse members of the human community, will do better then, on other worlds—an implied hope of Stephen Hawking. It will also do better on Earth when it takes responsibility for itself as a whole, commits itself to care for the common ground that it shares with all life, and works to secure the common good for all, including a better life for the members of its communities most in need of justice, compassion, and respect.

Then, perhaps, ETI will not fly by, but drop by: to make Contact in an extended way, and to engage in congenial conversation and collaborative, community-oriented ventures with humans. We will have evolved sufficiently—biologically, socially, psychologically, intellectually, and spiritually—to become interrelated intelligent members of the cosmic community. We will be received, respected, and welcomed, then, by our ETI relatives throughout the cosmos.

7

SPIRIT, SCIENCE, AND SPACE

Who Are the "True Believers"?

WE LIVE BY OUR material wellbeing and by our beliefs. We safeguard our physical development and wellbeing through nourishment, clothing, shelter, and health care (from native peoples' healers through technologically sophisticated surgeons) to keep our body going; stimulate our intellectual development and wellbeing through education (formal and informal, familial, communal, and academic), to learn about the evolving biological and social worlds in which we live and to promote creative thinking in new directions; and we sustain our psychic wellbeing in a variety of ways that include art, music, meditation of diverse types, being outdoors in forests or on mountains, philosophical speculation, openness to transcending experiences in and through nature, and through direct spiritual encounters with the sacred. Our beliefs might be in such areas as social and material progress, harmony among peoples of diverse cultures, our future professional accomplishments, or a particular ideology, philosophy, or transcending spiritual reality.

Atheist and theist, all of us are believers—including about the reality or not of a spiritual dimension of existence. Atheists believe that there is nothing of the sort; theists believe that there is. People with either perspective cannot "prove" their position: if they could, those of the other perspective would be converted to what had been viewed before as a contradictory mode of thought; and, moreover, "proof" would mean that they would no longer need to "believe": the facts would have established beyond a doubt the veracity and accuracy of what had previously been believed to be the case.

There is no contradiction between *materiality* and *spirituality*; there is a conflict between *materialism* and *spirituality*. We are obviously material beings; some of us understand ourselves to be in relation with a Spirit. Materialism, like spirituality, expresses a mode of consciousness, a state of mind, concerning how we understand our *be*-ing on Earth and in the cosmos.

Materialism or Materiality?

We who read *Encountering ETI* are corporeal intelligent beings; we have materiality on Earth and in the cosmos. That is, we have a physical reality, we are material—not solely spiritual—entities. In our materiality we are embodied beings. A materialistic belief, by contrast, focuses on materialism: all that is, is some form or expression of matter. There is no transcending aspect of our—or any—existence. Materialism as an ideology rejects dimensions or aspects of reality that are distinct from what can be demonstrated to be "real" in the material world: everything must be subject to the laws and manifestations of the intertwined and sometimes interchangeable space-time-matter-energy universe.

To the most ardent atheist reductionist, the only reality is the material, corporeal, physical. In this way of thinking in which materialism-based ideas and ideologies are not tempered by other considerations, if something cannot be quantified, measured, or falsified, there's no evidence that it is true (such as an assertion, description of data, hypothesis or theory) or that it exists (such as an event, energy, or entity). This philosophy and position, of course, is very useful for the development of scientific data, theory, and experiments; one has to establish in some verifiable way the veracity (truthfulness) or at least the credibility (capable of being believed, if only as a working tool or temporary postulate) of something. But the materiality of something need not be the basis only of materialism, in which only matter-energy that is observable or at least discernible in some way deserves serious consideration because it has the potential to be proven and accepted, or disproven and dismissed (at least temporarily, if not permanently) for lack of evidence.

In past eras, religious faith and religious thought were rejected on this materialistic basis by some thinkers. These aspects of human life and culture are part of the *metaphysical* dimension of reality that is not approachable with tools, technologies, and theories that are useful and necessary in the *material* dimension of reality. The rejection of spirituality by science had a sound basis, in some cases, wherever culture, politics, economics, ethnocentrism, and similar societal influences played an obvious role in religious thought, teaching, and practices, and so religions' ideas and religions' leaders were used to reinforce particular social practices and ideologies. When the insightful and compelling ideas of Copernicus, Galileo, Darwin, and Einstein, among others, came on the intellectual scene, for some people the rationale for religious reasoning was no longer needed and ceased to be operative.

In the recent present, perhaps particularly because of the narratives regarding Roswell and widespread acceptance of statements made by U.S. government officials in extensive efforts to suppress stories of UFOs and ETI, it became especially in vogue, in some scientific circles, to put thinking about extraterrestrial life in the same category as religious faith. "Where's the evidence?" became the rallying cry—"evidence" meaning, as in their attitudes and approach to religious faith, something tangible, material, subject to verification or falsification by existing scientific theory or laboratory experiment. Testimony by witnesses—again, as with religious faith—was discounted as illusion or intentional deception (there were cases of both—but other cases where people firmly affirmed what had transpired and passed lie detector tests as they recounted their experiences). Such supposedly "evidence-based" thinking (ironically, parallel to fundamentalist Christians' efforts to hide their creationist faith behind a supposed scientific veneer of "intelligent design") at times masked an atheism-based faith, a troubled religious faith, a psychological fear, or a professional fear about losing employment. In any event, it helped defund U.S. government support for scientific endeavors involved in the search for extraterrestrial intelligence (such as the SETI Institute). The result? "Flying saucers" and "little green men" became popular pejorative terms that were used extensively to ridicule and amusedly dismiss stories about and the existence of extraterrestrial vehicles and voyagers. Historically, this type of tactic proved to be ultimately futile in a similar situation, when scientist Fred Hoyle disparagingly dismissed the priest-scientist Georges Lemaître's theory that a "primeval atom" had exploded and expanded on a "day without yesterday": Hoyle called it the "Big Bang" theory. Within just a decade or so after Hoyle's derogatory comment, other scientists established that Lemaître was right;

Spirit, Science, and Space

two of them won the Nobel Prize for their secondhand work. Today, Lemaître's theory is recognized as a creative scientific breakthrough, a foundation for numerous subsequent strands of physics, including quantum physics.

Similar attitudes toward extraterrestrial existence persist today, but to a lesser extent. Advances in astronomy, physics, and mathematics, and complementary developments in scientific theory and technological practice, have led to what might be a transitional attitude. The discovery of quanta by physicists, of habitable planets by astronomers, of water or water traces on Mars, and of the principally oxygen atmosphere and ice on Europa, a Jupiter moon discovered by Galileo, as well as the elaboration of mathematicians' theories about multiple dimensions of reality and physicists' theories about black holes and dark matter—all have helped make what is currently theorized and unseen more credible. The unseen and the unforeseen might become the seen at some point. Most scientists now accept that there are significant possibilities that life-enabling planets exist even if undiscovered as yet, that life exists on some of those planets and, for some scientists, that life has evolved and complexified on several planets to become intelligent, and might have had continually evolving intelligent beings for perhaps a billion years or more before Earth was even formed—who would have acquired over that lengthy time significant technological prowess. "Dark matter" is accepted as part of "reality" since, even though it is not readily evident, there are scientific expectations that it eventually will be, and credible scientists who claim that it is "real." However, UFOs and ETI are not similarly accepted despite professionally accomplished credible witnesses' reports and scientific analyses of residues and other physical evidence found around the world at various and diverse sites that indicate that UFO landings have, in fact, happened.

Before we leap to too many conclusions and race too far ahead, it should be acknowledged that most scientists who accept the possibility of life elsewhere in the cosmos currently believe that it will be microbial life—life at its earliest stages of development. They are therefore more concerned, at this moment in time, that humans venturing into space might inadvertently export, on their tools and machines, microorganisms that would negatively impact—harm, cause to mutate, or kill—unseen, developing life on other celestial bodies. (This concern was heightened when Mars rover Curiosity used a drill that was no longer sterile to bore into Martian rocks and soil in search of evidence of life.) A more chilling possibility for Earth biota is that a pathogenic (disease-causing, life-threatening) transfer of microorganisms might follow the reverse trajectory, and be transmitted to Earth by accident (the precautions to prevent such a transfer were extraordinary when the first "moon rocks" were brought to Earth and eventually put on display).

In light of the preceding, it will be interesting to see skeptics' reactions to undeniable evidence of life—and especially of technologically capable life—elsewhere in the universe. The existing materialistic belief will have to be rethought—that faith is permissible only when it relates to accepting what has been or will be scientifically established or made scientifically credible based on past criteria: "Yes, Virginia, there are intelligent extraterrestrials." In such an instance, even if religious faith does not become more credible, at least those who give credence to stories about experiences of extraterrestrials will not have to be closet "believers" or "knowers"—as some scientists in major research universities are now. There will not likely be a parallel development in attitudes toward religious faith, since it still will lack some form of material basis—and some scientists in major research universities who are religious, or even "spiritual" in a less vacuous sense than the common usage of the word might suggest, will continue to closet themselves, their UFO curiosity, and their creative thinking about cosmos matters.

Integral Being and Sacred Materiality

When religious faith is grounded in real-life experiences or the testimony of credible witnesses, and related to the fullness of human experience as corporeal and spiritual existence on Earth (or elsewhere in the cosmos, and for not-human intelligent species as well) there is no rejection of the material world. Perhaps what might characterize or come to characterize the perspective of believers grounded in creation is a *relational religious rationality*, permeated by a sense of the *sacral materiality* of the cosmos; this will establish or enhance believers' relationship to all that exists in the integral being of the multidimensional, multireality cosmos. Spiritual-material intelligent beings acknowledge that the immanent-transcendent Spirit permeates the cosmos, intimately intertwined and interrelated with all-that-is.

Spirit, Science, and Space

Materiality and Spirituality on Earth and in the Heavens

In their consideration of the implications of TI-ETI Contact, Christians might come to comprehend (and subsequently experience) the *sacred materiality* of Earth and the cosmic commons as a whole. Divine immanence is mediated by a mode of transparency present in creation, while simultaneously the goodness and intrinsic worth of all beings in creation are appreciated for what or who they are, not solely as mediators of the divine. Appreciation for the cosmic Presence of the creating, loving Spirit can evoke praise, gratitude, and moments of immersion in the Spirit's engagement with and embrace of all life—terrestrial and extraterrestrial. Along these lines, the U.S. Catholic bishops, in their pastoral letter *Renewing the Earth* (1991), describe a "sacramental universe—a world that discloses the Creator's presence by visible and invisible signs."

A perspective that the cosmos reveals divine immanence can provide a foundation to understand and accept the presence of ETI: the creating Spirit's creativity has initiated an eonic and extensive historical trajectory, providing the stardust seeds of cosmic biotic existence. Indirectly and possibly unintentionally, this perspective is evident in an interview that José Gabriel Funes, a Jesuit priest, astrophysicist, and Director of the Vatican Observatory, gave to *L'Osservatore Romano* in 2008. Funes stated that it is very probable that there are intelligent beings elsewhere in the cosmos, and that people might view and relate to them much as Francis interacted with Earth biota and Earth and cosmos abiota: Francis called all beings "sister" and "brother." Funes suggested that such a familial perspective and attitude would help people understand and accept intelligent life from other worlds. After all, ETI shares in common with humanity not only life but intelligence, and perhaps a spiritual awareness. We are all, TI and ETI, thinking stardust who share a common origin billions of years ago in the birth of the cosmos.

Former Canadian Minister of National Defence and longtime member of the Canadian Parliament, Paul Hellyer, considered (in a response to a query I posed in a 2013 email) how religious leaders might respond or react to evidence of ETI: "This is difficult to respond to because I don't know that there is any appropriate answer. The Vatican is on record that there are other [alien] species created by the same God as ours and, in fact, has had encounters with at least one of them. I believe that it was one of the benign and friendly ones, so that would be appropriate because they are inclined to be more spiritual as well. In my book *Light at the End of the*

Tunnel I say, 'There is a spiritual overtone to much of the Extraterrestrial activity. Reports from many sources indicate that at least some species of Star Visitors are much more spiritual than Earthlings, and more reverential toward the Creator, the Source, the One, Allah, the Great Spirit—God by whatever name He is known.' I was referring there to the species that are really on the side of Light. . . . [But] one or two species . . . may be on the other side. The bottom line, however, is that the human species should be aware of and prepare to cope with both."

Consideration of Contact with extraterrestrial intelligent life offers an opportunity for reflection on the complexity and diversity of creation—and on humans' role to responsibly relate to and interact with other biota not only in the interdependent Earth community, but wherever their space explorations take them.

Being Out of This World Materially and Spiritually

The reign of God, the Christian gospels say, is simultaneously both "not of this world" and "within" Jesus' followers. It is not limited to this world—solely immanent in creation—but also transcendent to it, since it is a mode of thought that transcends material limitations. It is both-and: immanent-transcendent. It presents in theory and practice a different consciousness and value system than what is operative in society. Its contemporary rendition, as our beliefs about Creator and cosmos have evolved, is that it is the *commonweal of the Creator* or the *cosmic commonweal*. Christians are to bear it in mind and in spirit when they live their own Way directly on a daily material basis, and experience the spiritual dimension of reality in rarer moments. This internally cohesive material-spiritual complementarity could become beneficially operative in "far-out" places of the cosmos, if humankind were to change its current consciousness and conduct as it travels out of this world materially, and when it becomes spiritual-material ("transcends" materiality) as it sets aside its strictly material existence in the world at sacred moments and experiences the holistic interrelated dimensions of reality.

Spirit and Science: Teilhard, Lemaître, and Peacocke

In the twentieth century, three priest-scientists (yes, *priest*-scientists: two Catholics and one Anglican) from three countries creatively sought to integrate religion and science. Like their counterparts in previous centuries, these eminent scientists, simultaneously deeply spiritual, found no "conflict between religion and science."

Pierre Teilhard de Chardin, SJ

Early in the twentieth century, Pierre Teilhard de Chardin, SJ (1881–1955), a French Jesuit priest-geologist-paleontologist, pioneered efforts to integrate Spirit and science. He was an ardent advocate of the theory of evolution and of its reconcilability with Christian faith and thought (more privately and less publicly as his position was met with alarm and persecution by Vatican and Jesuit officials, and eventually erupted into open conflict). His scientific journeys, both intellectual and geographical, led to an impressive body of writings, international lectures, and participation in significant scientific expeditions as a geologist. (The most thorough biographies of Teilhard are Claude Cuénot, *Teilhard de Chardin*, from which much of the data presented here is drawn, and Mary Lukas and Ellen Lukas, *Teilhard*.)

Teilhard was born in France to theologically conservative Catholic parents. In his childhood years he developed a strong appreciation for nature's beauty and sought permanence in nature; he wanted to find and believe in things that endured. After several disappointments when he saw seemingly permanent objects perish, he settled on rocks as the most enduring, and his specific interest in geology was born. This closeness to nature contributed to his construction of concepts later in life that integrated his understandings of God and creation.

Teilhard was ordained a priest in 1911. From 1912 to 1914, he pursued scientific research in Paris at the Institut Catholique, Collège de France, Paris Museum, and Sorbonne. In December 1914, he was called to military service to fight in World War I (clergy in France are not exempt from military duty). At his own request he became a stretcher bearer at the front lines, and served in a Moroccan regiment comprised principally of Muslim troops. During battle he ducked and weaved among hails of bullets to rescue wounded soldiers and retrieve bodies of the fallen. His comrades admired his courage under fire; his officers commended his heroism with medals, including the Croix de guerre, Médaille militaire, and Chevalier de la Légion d'honneur. Even while in the military he continued to write scientific articles, and he collected field specimens of prehumans whose remains he discovered on battlefield sites on which artillery shell explosions had exposed subsurface artifacts. The war had a profound impact on him as he pondered life's deeper meaning amid the carnage around him. After the war, he professed his Jesuit vows in 1918.

Teilhard renewed his science studies in Paris, and received his doctorate in geology in 1922. As a Jesuit priest and geologist, he integrated his studies in theology and science (particularly geology and biology) and developed a dynamic theory of *cosmogenesis* about the origins of the cosmos and its apparent trajectory toward fulfillment in ultimate Being.

In 1923-24, he went on his first expedition to China, to Tientsin, where he took part in site excavations. He returned to Paris in 1924 and remained there until 1926. While in Paris, his sermons and his lectures at the Institut Catholique aroused Vatican and Jesuit interest and displeasure—but were popular among seminarians, parishioners, and the general public. The latter were interested in the relationship of evolution to Catholic dogma, particularly to doctrines concerning a "fall" of Adam and Eve and "original sin," and he addressed these topics as no one else dared do. His Jesuit superiors ordered him not to speak or write on theological subjects, only on scientific matters. He was forced to resign from the Institut Catholique and continue his exile in Tientsin. He worked with China's Geological Service and took part in major excavations at Choukoutien. In 1929 he was the geologist on the expedition that found an adult skull, tools, and evidence of fire. These discoveries indicated that *Sinanthropus pekinensis* was *Homo faber*, a tool-making ancestor of humans. Peking Man was, at the time, the oldest human remains that had been found to date.

His scientific work reinforced Teilhard's affirmation of evolution, and his dissatisfaction with the rigidity of the Catholic hierarchy and official Church teachings. Despite experiencing adversity within his Jesuit order and the Catholic Church, he did not break with either institution. He continued to hope that both would come to recognize ever-emerging scientific evidence for evolution, including human evolution, and appreciate his efforts to reconcile Church doctrine and scientific developments. He became resigned to exile, and plunged into his work. Although he was a priest, devoted Catholic, and developing Christian mystic, because of his religion-science ideas the Catholic Church kept him exiled from his French homeland and isolated him even from his family—he was not allowed to return to France on the occasions of the deaths of his parents and three siblings. The Church alienated him to China, and he became an alien in a land foreign to him. However, his situation enabled him to develop in greater depth his integration of religion and science. He formulated a vision for the future; he lamented that many people preferred to look backward rather than forward.

Living as an alien in China, Teilhard had opportunities to extend his scientific explorations in biology and paleontology. A fortuitous result of his exile—unanticipated by Church leaders—was the invitation he received to be part of the team of scientists searching for human origins that discovered *Sinanthropus pekinensis*. Ironically, Teilhard had been exiled because of his advocacy of evolution, and as a result of and during his exile he found further, significant evidence of human and other biotic evolution. His in-depth reflection on paleontological data regarding human origins and their relation to biotic evolution was stimulated and strengthened consequent to the discovery of Peking Man. He continued, too, to reconcile evolutionary theory with both Catholic doctrine, including biblical texts, and his experiential sense of a Creator God who was both transcendent to and immanent in creation. He died in exile while on a visit to New York in 1955.

Teilhard's ideas on science-religion compatibility, complementarity, engagement, and integration, particularly as expressed in evolutionary theory and data that he related to official church dogmas and his own scientific and religious intuitions and mystical experiences, were published posthumously. He had given or bequeathed his book manuscripts and essays to friends to ensure that they did not "disappear." His writings included several works that were science-religion themed, integrating evolution and Christian doctrine, and would have been denied publication by Rome and the Jesuits. *The Phenomenon of Man* is a scientific book focused on human origins and evolution, and future possibilities for humankind in a dynamic cosmos in which divine presence is immanent and influential. It is an attempt to reconcile and integrate scientific theories and data with theological doctrine and ethics. The manuscript he had submitted earlier to Roman and Jesuit censors for Church approval had not been allowed public dissemination and publication. *The Divine Milieu* is a collection of essays that reflect on ways in which creation is permeated by divine presence. *Christianity and Evolution*, another essay collection, focuses on the compatibility and integration of Christian doctrine and evolutionary theory; the book includes an essay exploring the likely existence of evolved extraterrestrial intelligent life.

Teilhard's pioneering work on biotic origins and evolution on Earth is complemented by his fellow priest-scientist George Lemaître's work on abiotic origins and dynamics in the cosmos.

Georges Lemaître

Georges Lemaître (1894–1966), like Teilhard de Chardin, was a priest-scientist whose ideas promoted the independence and compatibility of science and spirituality. J. J. O'Connor and E. F. Robertson, in their 2008 online article "Georges Henri-Joseph Edouard Lemaître," cite a *New York Times* article published in 1933 that featured a photo of Einstein and Lemaître beneath which was the caption, "They have a profound respect and admiration for each other." The author of the article observed that "'there is no conflict between religion and science,' Lemaître has been telling audiences over and over again in this country. . . . His view is interesting and important not because he is a Catholic priest, not because he is one of the leading mathematical physicists of our time, but because he is both."

Lemaître was born in Charleroi, Belgium. When he was seventeen he enrolled in the school of engineering at the Catholic University of Louvain (UCL); he left three years later to volunteer in the Belgian army to fight in World War I as an artillery officer and engineer. His heroism under fire earned him accolades and several medals, including the Croix de guerre avec Palme. By war's end he had lost interest in an engineering career. His postwar studies shifted from engineering to mathematics at the UCL; he received his PhD in mathematics in 1920. He was ordained in 1923; unlike Teilhard he did not belong to a Catholic religious order.

After ordination, Lemaître studied at Cambridge University with noted astronomer Arthur Stanley Eddington, who initiated his studies in contemporary stellar astronomy and related mathematics. In 1924, he studied at the Harvard College Observatory under the direction of another renowned astronomer, Harlow Shapley, noted for his work on nebulae. He received his second PhD, in physics, at the Massachusetts Institute of Technology, in 1927. The same year, he published in the *Annales de la Société scientifique de Bruxelles* an article that would establish his reputation later, when this seminal scientific treatise was more widely disseminated: "A homogenous Universe with constant mass and increasing radius explaining the radial velocity of extragalactic nebulae." In the article, he elaborated his groundbreaking concept of the physical expansion of the universe. He would call his theory of a singular, explosive origin of the now expanding universe "the day without yesterday." In his "primeval atom hypothesis," he stated that from a singular explosive burst of a quantum the entire universe eventually emerged, and has continued to expand.

When Lemaître discussed his article with Einstein at the Solvay Conference of physicists in Brussels, Einstein told him that "your calculations are correct, but your grasp of physics is abominable"—principally because Lemaître had used Einstein's theory of a cosmological constant, which Einstein had discarded, and proposed an expanding universe, which Einstein had rejected in favor of a steady state universe. In 1929, astronomer Edwin Hubble discovered and published astronomical data that seemed to provide evidence of an expanding universe, unaware of Lemaître's published paper. Lemaître sent his paper to Eddington, who had it translated into English and published. His ideas were at first ridiculed by those who thought that as a priest he was trying to use scientific theories to justify Genesis 1, but when astronomical observations reinforced his ideas they became more widely accepted and disseminated. In 1931 he explained his core ideas in an article in *Nature* in which he stated that "if the world has begun with a single quantum, the notions of space and time would altogether fail to have any meaning at the beginning; they would only begin to have a sensible meaning when the original quantum had been divided into a sufficient number of quanta. If this suggestion is correct, the beginning of the world happened a little before the beginning of space and time."

Lemaître and Einstein kept their now congenial conversations going for several years, often in person and trailed by reporters when they were concurrently in residence at the Institute for Advanced Study, Princeton, New Jersey. In 1934, he received from King Leopold III Belgium's highest scientific award, the Francqui Prize; in 1951, he was selected by the Royal Astronomical Society to be the first recipient of the Eddington Medal. Shortly before he died in Louvain in 1966, Lemaître learned that Bell Laboratories physicists Arno Penzias and Robert Wilson accidentally had discovered background interference when doing radiometer tests studying stellar radio sources. They had been contacted at the time by Princeton scientists Robert H. Dicke and P. J. E. Peebles, who with their team had done research on background radiation that should have resulted from a fireball origin of the universe, and had predicted that this "cosmic microwave background radiation" would be found. This was, of course, the radiation residue that Lemaître had theorized would be in space in the aftermath of the primeval atom "fireworks." In 1978, Penzias and Wilson received the Nobel Prize in physics for their accidental "discovery," whose significance they would not have understood without the work of Lemaître, Dicke, and Peebles. Papal activism in publicly linking Genesis 1 stories to "the day without yesterday" and the fact that Lemaître was a Catholic priest might well have been the principal reasons why he did not receive a Nobel Prize for his creative, insightful theory.

Most of Lemaître's scientist contemporaries initially dismissed and even disparaged his theory about a "day without yesterday" because he was a Catholic priest and because they thought that he was trying to provide a scientific justification for the Genesis 1 creation story. In a 1950 radio program, Fred Hoyle declared that Lemaître's idea was nothing more than a "Big Bang" theory; that dismissive designation continues today, although now as a popular phrase it is used positively to indicate approximately what Lemaître had proposed. Lemaître always made a careful and even sharp distinction between his theological beliefs and his scientific theories, to the extent of pointedly asking Catholic Church officials who linked publicly his scientific accomplishments and Genesis narratives to cease doing so. He advocated autonomy for both of the "two ways," as he called them: science and revelation.

John Farrell, in his biography of Lemaître, *The Day Without Yesterday: Lemaitre, Einstein, and the Birth of Modern Cosmology*, reports that in a 1933 interview, when asked how he could reconcile faith and physics, Lemaître responded,

> The writers of the Bible were illuminated more or less—some more than others—on the question of salvation. On other questions they were as wise or as ignorant as their generation. Hence it is utterly unimportant that errors of historic or scientific fact should be found in the Bible, especially if errors relate to events that were not directly observed by those who wrote about them.... The idea that because they were right in their doctrine of immortality and salvation they must also be right on all other subjects is simply the fallacy of people who have an incomplete understanding of why the Bible was given to us at all.

Farrell notes further that when asked in an interview decades later if the Big Bang theory was inspired by his religious faith, Lemaître's reply complemented the preceding response: "As far as I can see, such a theory remains entirely outside any metaphysical or religious question. It leaves the materialist free to deny any transcendental Being. He may keep, for the bottom of space-time, the same attitude of mind he has been able to adopt for events occurring in non-singular places in space-time. . . . Science has not to surrender."

The ideas expressed by Lemaître in physics and cosmology complement—indeed, overlap with—those of Teilhard in geology and biology. Their respective significant and science-impacting discoveries occurred just two years apart: Lemaître published his paper on primeval atom theory in 1927; Teilhard discovered Peking Man in 1929. Teilhard and Lemaître understood well, as scientists and Catholic priests, that there is no incompatibility between science and religion—even if the Catholic Church did not accept that fact (at least publicly). By the mid-twentieth century the priest-scientist contemporaries had experienced contradictory receptions from the faith community and the scientific community. Teilhard de Chardin had been accepted by scientists and rejected by the Catholic Church because of his scientific work; Georges Lemaître had been accepted by the Church because of his scientific work and rejected by numerous scientists because of his religious faith. Both priest-scientists ultimately influenced—and continue to influence—the fields of religion and science, which they pondered in depth and about which they spoke and wrote passionately.

Arthur Peacocke

The biochemist-Anglican priest and theologian Arthur Peacocke (1924–2006) was a pioneer in DNA research in the U.K., Dean of Cambridge University, Professor of Chemistry and Professor of Theology simultaneously at Oxford University, founder of the Society of Ordained Scientists, a member of the Most Excellent Order of the British Empire (appointed by Queen Elizabeth), and recipient of the 2001 Templeton Prize. He is noted as an accomplished scientist, an eminent theologian, and a perceptive and creative voice in the science-religion field.

Peacocke explores divine and cosmic dynamics and biotic evolution in several books, including *Theology for a Scientific Age: Being and Becoming—Natural, Divine and Human*; *Paths from Science towards God: The End of All Our Exploring*; *Evolution: The Disguised Friend of Faith?*; and his final book, *All That Is—A Naturalistic Faith for the Twenty-First Century*, completed days before his death.

When he addresses the so-called science-religion conflict in *Paths*, Peacocke states, "New voices have been heard from within the community of science" that "challenge dismissive attitudes towards religion and theology which are supposedly based on science.... The voice of science itself is not accurately represented by the anti-theistic guru-scientists" who attack religion and theology. Scientists' "commitment to excellence in science [is] clearly not for them inconsistent with commitment to religion—even to highly specific traditions of belief and practice.... For them, the scientific and religious quests [are] explorations into realities—two vocations that are intertwined, indivisible and mutually sustaining."

After lamenting how humans are destroying their planet, and rather should take responsibility for their world, Peacocke muses in the epilogue to *Paths*, "Might it not be, after all, that even modern (yes, and postmodern) humanity must come to recognise that the Reality that encompasses us is in fact the Source of all our existence and is the End of all our exploring for fruition, and so that to which we have willingly to adapt?" People must care for creation and be open to the Spirit of the Creator who is its Source, the One for whom people yearn in their quest for meaning in life—whether or not they recognize or acknowledge this.

When discussing divine interaction with the natural world, Peacocke states in one of his last books, *Evolution: The Disguised Friend of Faith?*, "The world appears to us less and less to possess the predictability that has been the presupposition of much theological reflection on God's interaction with the world since Newton. We now observe it to possess a degree of openness and flexibility within a lawlike framework, so that certain developments are genuinely unpredictable *by us* on the basis of any conceivable science."

Peacocke's continuing quest is to reconcile aspects of Christian faith with his considerable scientific knowledge. His work is permeated by an enduring faith in God-creating. There can be no incompatibility between science and religion, he states, since science investigates aspects of God's creativity (doing so, as is appropriate, without acknowledging any spiritual dimension of its work) that are seen in cosmic dynamics and biological evolution. God is the origin of the singularity with which the universe began; the Creator imparted creativity and freedom to creation. Peacocke writes, then, as someone with a deep theistic faith who wants to discern how Christian doctrines can be reconciled with science.

In Arthur Peacocke's thought, when scientific data are established, and compelling theories are proposed, Christian doctrine and biblical verses

Spirit, Science, and Space

and their interpretation must adapt to scientific facts and theories. His ideas on Jesus' incarnation, conception, and resurrection are not "orthodox," that is, compatible with traditional Christian beliefs as currently understood and elaborated. However, like St. Augustine, he affirms that if there is a contradiction between the Bible and science, the Bible must be interpreted allegorically. Peacocke thinks that in a scientific age it is important, in order that people might be more receptive to Christian thought, that science and Christianity be reconciled intellectually, to the greatest extent possible without compromising the integrity of either science or religion.

Spirit versus Science: Intelligent Design, Imperfect Design, Autopoietic Design-in-Process

In current scientific and theological thinking about how biota have emerged and evolved, the word *design* is viewed as an inaccurate description of the processes that are unfolding in creation. Analysis of evolutionary processes and results reveals no divinely ordained design in nature. Theists disagree among themselves about God's creativity as revealed in biotic development, and as to the manner in which biota began and have continued to exist; theists and atheists disagree about the meaning of autopoeisis in cosmic creation and biotic evolution.

Creation versus "Intelligent Design"

As discussed earlier (chapter 5), toward the end of the twentieth century some conservative Christians developed a theory—known as intelligent design (ID)—which stated that the universe was intelligently designed. They did not specifically say that God was the designer because previously U.S. courts, concerned to continue the traditional "separation of church and state" in the U.S., had rejected creationists' claims that creationism was "creation science" and dismissed creationists' efforts to force public school systems to include the religious doctrine of "creationism" in the curricula of biology courses. The unscientific religious doctrine of creationism taught that the Bible's stories, intended by biblical authors to teach religious beliefs and moral values, were, effectively, less than sacred texts with religious meanings: they were secular texts, science and history books that were literally "true." Such "truth," which creationists/ID advocates claimed to be scientifically and historically accurate, included doctrines that the Earth was a few thousand years old; that God had specifically created all species on Earth, and therefore biotic evolution had not occurred; and that humans and dinosaurs walked Earth at the same time. To develop these doctrines, creationists had not used *exegesis*—thoughtful analysis to determine the meaning of specific biblical texts, including by studying their original historical context in a particular culture, the available prescientific data that the writers used (based primarily on what the writers could see or touch), and the message the writers intended—but rather *eisegesis*—imposition of creationists' present beliefs onto ancient texts, written two millennia or more in the past, to give them a meaning and a function not intended by the original authors. Rather than seek to understand the biblical writers' message, creationists forced their own religious beliefs and message onto the biblical writers' verses.

The ID proposal presents creation as a *machina ex Dei*. It embodies, too, a certain *mathematica ex Dei*, reducing creative freedom—God's, intelligent beings,' and the universe's—to mathematical formulas in an effort to put a scientific "face" or veneer on "creation science," itself an expression of literal interpretations of Bible texts; it is biblically related (not biblically based) creationism. In ID thought, creation is not seen as the dynamic, creative, free integral being that it is, but as the predetermined development of a rigid divine plan, under initial and ongoing divine control. ID proponents expected their belief to be accepted as a legitimate "science" that should be taught as an option in U.S. public school science classrooms, on a par with the natural sciences, whenever cosmogenesis and cosmic dynamics, and biogenesis and evolution, are presented in science courses. Their quest was thwarted in federal court, where judges detected and rejected ID's fundamentally religious, not scientific, nature.

ID is not *cosmology*; it is similar to *cosmetology*. In beauty salons, the beautician uses cosmetics to alter the appearance of their client, and thereby to alter to some extent that person's inherent sense of who they are, or inherent projection of whom they would like to be or at least appear to be; it is a human imposition on existing materiality, physiology, and psychology. In like manner, the ID advocate seeks to shape humans' perception of the materiality and energies of continuing, dynamic creation according to particular limited and limiting current religious-cultural human conceptions that flow from a particular religious base and rigid dogmas. ID strives to put a cosmetic "mask" or human "face" on divine creation to make the cosmos appear to be that which it is not. ID masks creation's nature, its dynamic, divinely imparted creative freedom, under an imposed cosmetic design—just as is done to a woman or man in a beauty salon on Earth, but on a much larger scale. ID hides, beneath a human-constructed and human-satisfying veneer, what has naturally been formed and continues to be forming in creation.

Creationism in itself and as ID is not wholly erroneous; it is a belief system whose adherents have a right to hold and to seek to convince others to embrace. While not scientific (and with anti-science theories and beliefs), and while it denies that God gave freedom to creation to develop cosmic dynamics and biotic evolution, at least (from a theist point of view) it teaches that divine creativity is immanent in the cosmos. ID attempts to limit that creativity, however, to a specific formulation expressed in Genesis creation stories that were not intended to narrate history or science, written in an era in which the ancient Hebrews had no scientific instruments to study astronomy and had a limited understanding of the cosmos. All materiality that exists was, for the biblical writers, entirely contained in Earth and Earth's atmosphere, which included a "firmament," a hard ceiling that separated waters above and waters below, a firm structure to which heavenly bodies were attached and on which they followed their assigned tracks to orbit Earth. Not even the most hardened, convinced creationist today states that such a dome is atop Earth and that stars are lights attached to it that follow tracks that circle Earth and sustain stellar orbits.

Contrary to what ID advocates profess, the Creator Spirit does not premeditatedly specify a prescribed "design" that predetermines all that comes to exist after the singular "day without yesterday" and coerces all being—abiotic and biotic—to follow the design established for them. Rather, as revealed unintentionally by scientific discoveries, whose focus is materiality, not metaphysics, the loving and creatively free intelligent Spirit initiates, in giving birth to origin existents from divine Being, a mode of creative freedom that enables abiotic cosmic dynamics and biotic evolution. In reflecting on the process of biotic evolution, after observing that "natural selection primarily promotes adaptation rather than evolution," evolutionary biologist Francisco J. Ayala, in *Darwin and Intelligent Design*, states, "Natural selection is simply a consequence of the differential survival and reproduction of living beings. It has some appearance of purposefulness because it is conditioned by the environment: which organisms survive and reproduce more effectively depends on which variations they happen to possess that are useful in the place and at the time the organisms live.

But natural selection does not anticipate the environments of the future; drastic environmental changes may be insuperable to organisms that were previously thriving. Species extinction is a common outcome of the evolutionary process." In other words, environmental changes such as dramatic or even drastic climate change can alienate species from their customary abiotic context, and ecological changes such as an invading alien species that compete with them in their biotic relationships within their context can alienate them from essential water or nutrients. Only species—or individual members thereof—that have genes that can adapt to or mutate into the new environmental or ecological setting will survive and evolve further—and be, effectively, a species or subspecies that is alien even to their own ancestors.

Ayala and numerous other scientists, as well as theologians in mainstream Christianity, have pointed out time and again a fundamental contradiction: if God is the "Designer," why is the "design" imperfect? The human eye is one example of imperfection. Humans need eyes in order to have a good view of their surroundings and adjust to them as necessary. But the eye has to modify what it sees to make it conform to reality. In the curious mechanics of the eye, for example, when people view an object, the image on the retina is inverted (upside down) and reversed (from right to left) rather than appearing as it exists in the world in itself, external to the eye. The human brain corrects the imaging error. The octopus eye, by contrast, is exceptional and serves the octopus well deep beneath the surface of Earth's oceans. Would imperfect design of complex biota, as happened to humans and others, come from the perfect, divine designer in whom ID proponents (and others) believe?

Intelligent design is a belief system and not a science. It is best taught within ID advocates' own conservative Christian places of worship and teaching facilities—just as is done by members of other religious faiths, who convey their beliefs and doctrines in their dedicated religious buildings and educational institutions. As is the case with doctrines and stories from other religions (which have their own diverse creation stories and different sacred books), ID should not be imposed by governments or public education institutions on those who do not choose to accept ID's particular doctrinal base, or its perspective on divine action and creativity.

As humankind continues to adapt and evolve, having reached a stage of complexity previously unknown among Earth's biota, ever-increasing knowledge, ever-more complex hypotheses, and ever-higher stages of consciousness emerge. Think about all the information available today on a computer, when a few keystrokes provide access to near and distant sites and enable online accessibility to recognized experts on a broad range of topics. (It's all a student's dream for a research paper or doctoral dissertation!)

Humans have evolved from "the realm of necessity to the realm of freedom" (to adapt a phrase Karl Marx uses for economic equality progress) in that we have the capability to responsibly locate and use Earth's natural goods and places to provide for humans' (and other biota's) needs in a socially and ecologically responsible way. The acquisition of life's necessities has become for many people a given; it is no longer the ongoing, vital struggle for survival as had been the case for species and individuals in former times and places. However, this is not the case for the majority of people: vast disparities exist still between rich and poor, in terms of both the acquisition and use of Earth's goods, because the greed, control, and avarice of the few operate at the expense of meeting the basic needs of the many.

Evolution from Theistic Faith to Atheistic Faith?

Some atheist or humanist theorists believe that humankind—or at least some people—had an evolutionary progression from a "necessity to believe" as a survival mechanism in human consciousness, to a stage where religious faith is no longer necessary and can be dispensed with. In thus seeking to establish religion and spirituality as aspects of a distinct (if lengthy) and solely temporary phenomenon entirely separate from an extended, common historical moment of shared evolutionary materiality, such theorists declare that as humans increase in knowledge and levels of intelligence far superior to those of their ape ancestors, their "maturity" will enable—indeed, prompt—them to go beyond religious-spiritual beliefs and consciousness. All humankind (or at least the "most intelligent humans"), the atheist humanists believe and theorize, will inevitably evolve into a superior state of intellectual and social acceptance of their materiality: they will not need religions' support for societal morality, social stability, and spiritual security. Such theorists believe that materiality is the fundamental nature of all life; the smarter, more intellectually advanced people and species that will evolve will not "need" spirituality to provide meaning and intellectual comfort in life. Currently, many in this intellectual materialist elite are fond of saying that one cannot be both a scientist and a Christian believer (past and present people who are simultaneously both notwithstanding). A scientist who goes to church must "leave their brains at the churchhouse door" prior to entering to worship. This attitude has become pervasive enough that research and teaching scholars who are members of science departments keep their religious affiliation secret, fearful of being passed over for a promotion or a needed salary increase, or even losing their job. Georges Lemaître's reception by many members of the scientific community is akin to what scientists who are Christians experience today at the hands of atheist scientists who go beyond their scientific expertise. As they transcend their work in scientific fields in which they study the material (matter-energy) dimensions of reality, they venture into metaphysics and the spiritual dimensions of reality without the professional competence to do so, or the personal commitment to explore these dimensions openly and in depth. Such atheist scientists rightly dismiss fundamentalists' interpretations of science, which are beyond fundamentalist religionists' area of competence, and creationists' advocacy of a nonscientific (and nonbiblical) theory of intelligent design. But they commit the same type of error as the creationists when they speculate metaphysically (even while claiming that theirs is a material speculation) about spiritual dimensions of reality.

Today, aspects of the prevailing scientific theory that the cosmos began from a singularity, a unique state of being whose prior origin is unknown, continue to be debated: is the singularity a result of divine creativity or cosmic dynamics?

Creation and Imperfect Design

Francisco Ayala suggests that ID stands for imperfect design when the processes of creation, evident in evolution, are studied by scientists. Ayala discusses several biotic physical "imperfections" and "defects" that would not be present if God had designed all creatures; God would have done better than what creationists' claim God did in creation. The "designer" of biota is, so to speak, evolution by natural selection. In fits and starts evolution unfolds as biota seek a niche, establish themselves, and then adapt or go extinct as conditions change. Science, not intelligent design, provides explanations for defects, deformities, and dysfunctions that emerge during biotic evolution. Evolution, not intelligent design, enables organisms to use new combinations of DNA or segments of previously unused DNA to adapt to Earth changes. Biota gradually complexify when helpful, advantageous new combinations arise and increase, during which time they are incorporated internally or externally. Existing or new combinations that are disadvantageous are, effectively, examples of imperfect design.

Creation and Autopoietic Design-in-Process

"Creation" implies a Creator, however this Being might be understood in diverse cultures and historical eras. While some theists claim that the Creator has a "design" for the universe, others assert that the Creator imparts creativity and freedom to the cosmos and to intelligent life made in the Creator's "image and likeness" not physically (since God has no corporeality) but to express on Earth in visible form God's love and compassion, responsible freedom and creativity, and care for and in creation.

Autopoeisis is "self-organization" or "self-design." Atheist scientists often advocate autopoeisis as an alternative to a creation by a transcendent Being. Secular humanists believe that when the cosmos emerged in an explosive Big Bang, no Creator was involved. Creationists, in turn, reject scientific data about the age of the cosmos and of biotic evolution. They believe that all was created by God about six thousand years ago; that all species were created by God to dwell on Earth—no evolution was needed for biota to arrive at their present biological state or environmental locale; and humans are a "special creation."

Autopoeisis has distinct meanings for theists and atheists; creationists would accept none of them. Absent creationist beliefs, theists and atheists alike have a materialist understanding of autopoeisis—in ways that present a notable contrast. Science-informed theists accept autopoeisis as a viable and useful concept to understand how the cosmos unfolded and expanded or inflated *after* the singularity that originated in and from the creating Spirit. Science-informed atheists, as noted, see no Spirit influence on what they believe to be an entirely autonomous origin and development of the cosmos.

A material-metaphysical (material-spiritual) understanding of autopoeisis is that Spirit initiated evolutionary dynamics (at least in seminal form) that would continue through creation: the entities, energies, and fundamental laws of the cosmos. These were actuated at creation or afterward when complex combinations of contexts and space-time-element-energy factors catalyzed the emergence of latent cosmic capabilities over eons of time.

In the primordial cosmic moment Spirit, whose principal attributes include love, creativity, and freedom, created from love and imparted a spark of divinity in creation, in the form of freedom for autopoeitic creativity. Divine love, freedom, and creativity permeate and are part of the integral being of the dynamic cosmos. In the Christian Scriptures, Paul's letter to the Philippians teaches that God "emptied" God's self when becoming incarnate on Earth (Phil 2:6–7), setting aside divine power so as to experience, and influence without coercion, human consciousness and conduct. In like manner, God "emptied" God's self at the singular, initial moment of creation: rather than control or design what would unfold cosmically, the creating Spirit imparted creative freedom to a cosmos existing in the context of love, the primary element and energy of the Spirit. Within this creative process, influenced by parameters of being, autopoeisis occurs without divine control. Rather, the Creator might on occasion and because of the primacy of love in Creator and cosmos, stimulate the latent potential of all-that-is (or a part thereof) in order that it might become all-that-it-could-be.

ETI: Theist or Atheist?

In some scientists' thinking and judgment, just as highly intelligent humans no longer "need" religion, neither will intelligent ETI—if they exist. ETI would have biologically, intellectually, and socially evolved beyond such a belief. These scientists anthropocentrically and anthropomorphically extrapolate from what they believe about TI on Earth to what they state will be true about ETI in the cosmos. Their attitudes and consciousness are expressed in the real academic world when their belief becomes one of their criteria as they assess whether or not potential faculty should be hired or existing faculty should be promoted. In terms of both religious belief and ETI "belief," these atheist scientists who believe themselves to be "intellectually superior" assert that what offending scientists express regarding a Spirit or aliens is a "belief" without foundation, and therefore is irrational. Such "beliefs," they claim, demonstrate lack of intellectual ability since offenders have not reached the "intellectual maturity" stage at which they will no longer need a "crutch" of religious faith or a "belief" in anything else, for that matter. Theistic believers are regarded as those human beings—or ETI beings—whose evolution has been retarded or curtailed because their intellectual growth has been stunted.

In such thinking, ETI will have no religion or even spirituality (in its traditional meaning of relationship with a Spirit). In order to have attained their scientific and technological capabilities, they will have had to overcome the cumbersome weight of "unscientific" religious-spiritual thinking that impedes social and scientific progress. In recent decades, as speculation increased about the possibility of intelligent extraterrestrial beings, several scientists have expressed this view.

A sharply distinct and contrasting possibility exists: that *spirituality* is a higher stage of evolution than basic *materiality*. In this case, humans' planet-wide spiritual consciousness and corresponding conduct could become, upon Contact with intelligent species from other worlds, a continuing cosmos-expanding and cosmos-extensive spiritual consciousness, with corresponding conduct. It might find expression outside of formal religious institutions with their required dogmas, clergy, and rigid rituals. TI and ETI both will have instead an integrated, loosely associated ecumenical religious faith without fundamentalism, open not only to its own planetary evolution and development but to the spiritual consciousness and practices of other cultures that it encounters.

Human beings tend to extrapolate and extend from their own experience what will be the ideals, social constructs, and ideologies of other intelligent beings. Thus, atheists who claim that atheism is "intellectually superior," a "higher form of intelligence" that needs no spiritual understanding, and an "advanced evolutionary stage," project their belief onto ETI. Similarly, theists who see their religious faith and spirituality as a complement to their materiality project their belief onto ETI. Note that in both cases there is a fundamental belief, but neither group has evidence to support its particular perspective. Quite possibly, each perspective has the same odds of being accurate as does the other. Until longer-term Contact takes place in which a free exchange of ideas and beliefs occurs, both projections remain as possibilities; either is likely to be an ETI worldview or even ETI *worldsview*. ETI might have other possibilities, too, which we humans have not conceived. And, there might not be a literal "universality" of intelligent beings' beliefs: it is possible that some worlds' inhabitants have evolved to an advanced, materiality-originating spiritual consciousness, and some have remained focused on and fixed in evolution toward a more complex material consciousness.

One aspect of this discussion about which both theists and atheists can agree, I think, is that in order to become one cosmic family with multiple diverse relatives of distinct and numerous cultures, ideologies, ethnicities, spiritualities, and species, we must learn not just to "tolerate" different others but to acknowledge, respect, accept, and embrace them as family members, and as our neighbors on Earth, on other planets, and on celestial bodies. In Jesus' phrasing, we must love our neighbors (all other people) as ourselves, in accord with the Great Commandment to love God and love our neighbor. Our "neighbors" live not solely next door or on our home planet, but throughout space.

On Earth it is difficult to foster globally mutual respect while current disparities exist in economic wellbeing, and current political disputes are present and smoldering, or already have exploded into regional armed conflicts. We must go beyond recognizing that we are one species to interacting as if we are one family: our many local families, affiliated by kinship, must become a family of families globally. Disparate communities must become related communities and then a community of communities. We are, after all, "*evolved* stardust," striding from a common origin and striving toward the Garden utopia. We are evolved rising apes, not fallen angels. If we learn to accept this ourselves, we will be open to consider cosmic social, scientific, and spiritual understandings advanced by ETI and to integrate them to some extent with our own.

Spirit, Science, and Space

As humankind gradually transitions into the "Space Age" on its way toward the "Star Age," scientists and spiritually conscious people, in their respective ways, consider the implications of this next era of human existence. Along with wondering how all of this will affect the human psyche, they speculate how their respective scientific and spiritual views might be shared—or not—by ETI who are encountered on Earth and exoEarth.

Science and Space: Atheist Faith-Based Scientists and ETI

Atheist scientists who have faith in their assertions that there is no God, Spirit, or divine Being by any name, and declare that theists' faith would be threatened by the existence of ETI, might prove to be more—or equally—threatened by the possibility, let alone the actuality, not only that ETI exists but that ETI has an evolved, superior form of spirituality. As noted previously, many of these scientists are fond of stating that as intelligent biota

Spirit, Science, and Space

evolve biologically and culturally, they will not "need" religion or belief in a transcendent being. They claim further that theists would be deeply disturbed by Contact. This hypothesis might or might not be borne out in human historical experience. Studies have shown, to the contrary, that most theists are adaptable, and adopt their doctrines to incorporate aspects of materiality, to some extent, when confronted by irrefutable data that contradicts their long-held beliefs.

A complementary intellectual exploration would be an inquiry into the possibility that scientists' atheist faith would be particularly challenged if the reality of ETI were to be established definitively. No longer would people "believe" in ETI: since we believe in what is unseen, seeing is *not* believing: it is *more than* believing. To what extent might atheist scientists be amenable to contextual influences in terms of their belief in their intellectual superiority among humans, which they extend on a cosmic scale, or their related beliefs that ETI does not exist or that, if it does, it will have long since rejected religion during its biological and social evolution? What might happen if these cherished beliefs are proven to be unfounded? Some scientists have confidently assured themselves that Earth is the only planet on which life has emerged and evolved and that, even if biota do exist, they are not complex, intelligent biota. Thus, as atheists and Earth scientists, they are the "top dogs" in the cosmos.

Scientists who say, effectively, "seeing is believing," which is often expressed as "where's the proof?" might continue to erect intellectual barriers against even the most credible witnesses' reports about UFOs and ETI—such as data presented by astrophysicists from observatories around the world—and declare that they have to see material evidence "on the ground," figuratively and literally, in order to find these reports acceptable. Yet, they accept as true the assertions by U.S. government nonscientist public relations personnel that "there's no such thing as UFOs . . . or ETI." Why are scientists afraid to consider, or to boldly suggest, that this is government propaganda, prevarication, or disinformation?

Allen Hynek has suggested that some scientists might be afraid of being displaced as the cosmic "alpha male" by ETI. He speculated about this when, as a highly respected scientist who worked for the USAF/U.S. government, he interviewed credible witnesses during his UFO investigations. On at least two occasions, as a scientist who tried to evaluate objectively what he saw and heard and keep it "at arm's length," he became uneasy when he spoke with sincere people who obviously had been affected by what they'd seen. Their narratives—and the witnesses themselves—were very credible. Up to that point, Hynek had been able to maintain a certain scientific objectivity that kept data at a distance. Suddenly, he experienced its impacts directly, personally, and deeply, even without having seen phenomena akin to what witnesses had seen. What he learned from witnesses during his direct contact with them was troubling and unnerving. In this context, he wondered if other scientists, used to theorizing and explaining, would be similarly impacted upon realizing that ETI is smarter and more powerful than TI. Aliens would then at least psychically, if not physically, dis-place humans as the most intelligent of intelligent species.

In a way parallel to atheist scientist statements that theists would "lose their [religious] faith" if ETI were Contacted, it might be suggested that atheist scientists would "lose their [reductionist] faith" in their exceptional scientific expertise and their exceptionalist, elevated social status once ETI data became irrefutable even by the very high standards to which they hold religious belief and "belief" in UFOs and ETI. They often have applied scientific criteria to judge faith in a Spirit. However, ironically, they have faith—in predecessor scientists' data, claims, and theories—without establishing empirically by and for themselves, through controlled experiments, what past scientists in their field have stated is "true" instead of "credible." They do not replicate every lab experiment, even solely in their specific field for only one century, to establish "for themselves" that what they have accepted "on faith" is indeed a fact. They might state that there is evidence for predecessors' work that is visible in the material impacts that it has had in science and in society. They are unable to accept comparable evidence, on a metaphysical level, of societal and even scientific impacts presented by people—including scientists—who have experienced in some way the spiritual dimension of reality. They insist on applying to the *metaphysical* dimension the criteria that are needed and useful in the *material* dimension to judge claims about material aspects of reality.

Like apples and oranges, materiality and spirituality are distinct realities; also like apples and oranges, which are both fruits, they are interrelated realities. After all, material humans have metaphysical-spiritual moments. Material humans live their biotic lives in ways affected by metaphysical experiences—others' and their own—and their spiritual lives are impacted by their materiality. The material and metaphysical are inseparable for theists and other believers. All of us, atheists and theists, in some way "live by faith"—in our materialist beliefs, scientific studies, and self-confidence; in our spiritual beliefs, experiences, and relationship with the Spirit.

Spirit and Space: Indians' Insights

Indians' "Star People" Encounters across the U.S.

Indian peoples in the U.S. have numerous narratives about encounters with visitors from space. Over the years, I have heard directly from Indians about their experiences with ETI visitors, whom many call the "Star People." Interactions of this sort are presented, too, in *Encounters with Star People: Untold Stories of American Indians* (2012), by Dr. Ardy Sixkiller Clarke, retired Research Professor, and Professor Emeritus of Education, Montana State University. She presents narratives from diverse native peoples in the U.S., all of whom wanted anonymity for fear of ridicule in the Indian community and beyond—which would have impacted their professional standing as highly respected community leaders, educators, youth counselors, police officers, highway snowplow operators, and university faculty, among other occupations. Several spoke of experiences they had when in the military. Some traditional elders—spiritual leaders as well as civic leaders—related centuries of narratives about direct interactions with the Star People, including recent encounters. A few natives saw these visitors so often in childhood and adolescence—as often as three times per month—that they thought all people had such experiences because they were commonplace near their home. Most of the incidents occurred from 1990 to 2010; the interviews occurred as recently as 2012.

I have not discussed the issue of abductions previously in my writings because I understood that most stories related by people claiming to be abductees only emerged under hypnosis. However, Sixkiller Clarke presents the stories of several Indians—from university-educated to those with little formal education in schools—who experienced being taken into spacecraft. During their interviews with Sixkiller Clarke, *without hypnosis*, these credible witnesses recounted these incidents openly and directly. The abductees' stories, and other events described by Indians throughout the U.S., suggest that there might be four types of Star People, some of whom (with human features, but taller than most humans) are benevolent scientists studying Earth and biota, but others of whom ("insect-like"; "almost robotic"; without emotion) are malevolent: they kidnap people and subject them to experiments or medical examinations and procedures, and mutilate cattle and other animals in order to take body parts for examination or experimentation.

Star People is a remarkable, down-to-Earth book, even as it discusses exoEarth beings and ETI-TI encounters, because the interviewees are ordinary people who have seen or experienced extraordinary events and who relate them directly and simply. If their narratives are accurate, their stories resolve to some extent the debate between scientists—Stephen Hawking versus Peter Singer and Robert Wright, for example—about whether ETI will be benevolent or malevolent; the stories indicate: "both." Some elders state that the initial Star People who have been coming for millennia have been benevolent (some of whom are understood to have been Indians' ancestors). However, as signs of an intelligent human presence on Earth became detectable from space, malevolent aliens began to arrive.

Muskogee Indians' Encounter: "Boxes in the Sky" People

Phillip Deere (1926–85), a Muskogee spiritual leader, healer, human rights activist, and friend of mine, in 1984 recorded in my Montana home the migration story of his people. For more than three hours, without notes and without stopping, this native from Okemah, Oklahoma, narrated into a cassette recorder how the Muskogee had migrated on foot from the West Coast to the East Coast millennia or centuries ago after the Creator instructed them go on this journey. During the sixteenth to eighteenth centuries European explorers and traders came in contact with their communities and farms along waterways (thus they were called "Creek" Indians) in what is now the southeastern U.S. They were subsequently forced by the U.S. government to go back west to "Indian Territory," principally located in what is now the state of Oklahoma.

At one point when I entered the room as Phillip was relating the oral history, he reached a part that said that the Muskogee travelers met "people that came from boxes in the sky." Phillip paused in his storytelling, looked up at me, said nonchalantly, "Maybe they came from another planet," and returned to recording the story. He expressed neither concern nor fear about his ancestors' encounter with ETI; it was just another event along the way that had been judged sufficiently significant to be included and retained in the long Muskogee oral history. Since the aliens had not been aggressive there was no reason to fear them; since they did not, in the story,

Spirit, Science, and Space

seek an ongoing relationship with the Muskogee, there was no reason to dedicate to them a lengthier segment of the oral narrative. Perhaps, too, stories about a longer association were known and conveyed separately only orally, and so the brief phrase was needed to remind the Muskogee of events described in another narrative.

In the Lakota tradition, people acknowledge that all living creatures are "my relations." They greet each other and all life with the phrase *Mitakuye oyasin*—"we are all related." Spiritual leaders and other Lakota who address a gathering or open a meeting with a prayer will say, *How mitakuye oyasin*: "Greetings, all my relations." Some elders might elaborate: "Greetings to all the two-legged people. Greetings to all the four-legged people. Greetings to all the winged people. Greetings to all the finned people. Greetings to all the rooted people." It is not hard to imagine that, at Contact, elders and others would say to ETI arriving as aliens, *How mitakuye oyasin*, and add to the biota mentioned above: "Greetings to all the star people."

Star People and Paul Hellyer

In a remarkably candid 2013 public presentation, the Honorable Paul T. Hellyer, former Minister of National Defence of Canada, stated forthrightly that "UFOs are as real as the airplanes flying overhead." Hellyer has impressive credentials: he was an aeronautical engineer; during World War II he was employed at a facility that made training craft for the Royal Canadian Air Force, and later joined the Royal Canadian Artillery; he was a licensed private pilot; he was a Member of Parliament for twenty-three years and Minister of National Defence in the cabinet of Premier Lester B. Pearson; and he served on the Canadian Privy Council (whose membership included Prince Philip).

Hellyer presented his ETI ideas concisely during the Citizen Hearing on Disclosure Conference, held at the National Press Club, Washington, DC, April 29–May 3, 2013, a conference broadcast live on the Internet. He stated that as Minister of National Defence he had received numerous reports that extraterrestrial spacecraft and ETI had visited Earth, but he had no time to pursue the subject because he was involved in integrating all Canadian military branches into a single, unified force. His presentation is available in a brief, ten-minute YouTube video: "Breaking!!! UFO Alien Disclosure by Canadian Minister of Defence May 2013." (The reader seeking this and related videos is encouraged to search for "Canadian Defence Minister UFO.")

Hellyer's data regarding the number of alien species corresponds with that of Sixkiller Clarke in that he, too, states that "at least four species have been visiting Earth for thousands of years"; he adds that three of these species' names are known. The three have different agendas vis-à-vis Earth, which he does not specify in his presentation. He adds that representatives of two species who have been working with the U.S. government live on a U.S. Air Force base.

In his presentation, Hellyer addresses government assertions that people would be terrified on hearing that ETI exist. He states that "just as children survive the idea of the tooth fairy and Santa Claus when they become adult, I think the taxpaying citizens are quite capable of accepting the new and broader reality that we are living in a cosmos teeming with life of various sorts. The fact that some other civilizations are more advanced than we are may be humbling. It exists and is being kept secret." He demands government transparency on UFO/ETI files: "My interest is in full disclosure . . . at least 95 percent to 98 percent" of governments' UFO/ETI information "should be fully disclosed"; only "one or two things" should be withheld for national security reasons. Hellyer, in phrasing complementary to that expressed in this book, concludes that, for the wellbeing of Earth and all peoples and to be acceptable to ETI, "we have to become spiritual beings, and practice the one tenet that the world's religions have in common, that is, the Golden Rule."

In 2010, according to the Wikipedia article on Hellyer, he strongly disputed Stephen Hawking's statements that humans should avoid making Contact with ETI because such aliens likely destroyed their home planet and are seeking other worlds to devastate. Hellyer accused Hawking of spreading misinformation about who the aliens might be and what they intend to do in their space travels. He stated in an interview with Canadian media that "the reality is that they have been visiting earth for decades and probably millennia and have contributed considerably to our knowledge." He added that Hawking "is indulging in some pretty scary talk there that I would have hoped would not come from someone with such an established stature."

Spirit, Science, and Space: TI and ETI

In the works of twentieth-century thinkers Pierre Teilhard de Chardin and Arthur Peacocke, inklings of a possible integration of Spirit, science, and space are discernible. Their openness to the possibility of ETI makes them uniquely capable, along these lines, to provide seeds for theists to ponder ETI implications and to reconcile their beliefs, in their considerations, with the realities of ETI thought and action.

Teilhard: Extraterrestrial Worlds and ETI

In mid-1953, while in New York, Teilhard wrote "A Sequel to the Problem of Human Origins: The Plurality of Inhabited Worlds," which was published decades later in *Christianity and Evolution*. His scientific knowledge of biotic evolution on Earth (reinforced by paleontological data from his field expeditions) and of the age of Earth and the universe (reinforced by geological data from his expeditions) were supplemented by his understanding of the cosmic time required for Earth and life to have eventually come into existence. This inter-sciences linkage stimulated him to consider similar possibilities elsewhere in the cosmos. His ideas here were, as was the case with his thinking about the relationship of evolution and theology, well ahead of his time and place. They were written just six years after news of the crash of alien craft near Roswell, New Mexico, which had been followed by other UFO reports and scientific and theological speculations about extraterrestrial life possibilities—including in a 1952 *Time* magazine article, which he notes in passing.

Teilhard states in his essay that millions of galaxies exist in the universe, in each of which "matter has the same general composition and is going through essentially the same evolution" as happens in the Milky Way galaxy. Consequently, he speculates,

> If it is true that the proteins . . . appear in the universe as soon as it is possible for them to do so, and wherever it is possible,
>
> And *if*, when life has once taken hold on a star, . . . it carries itself as far and to as high a degree as possible (that is, up to "hominization" if it can),
>
> And *if*, in addition, there are thousands of millions of solar systems in the world in which life has equal chances of being born and becoming hominized,

> *Then*, our minds cannot resist the inevitable conclusion that were we, by chance, to possess plates that were sensitive to the specific radiation of the "noospheres" scattered throughout space, it would be *practically certain* that what we saw registered . . . would be a cloud of thinking stars.

This, he says, is "*the most probable* alternative" because "the idea of *a single* hominized *planet* in the universe" is "almost as *inconceivable* as that of a man who appeared with no genetic relationship to the rest of the earth's animal population." He estimates that there exists, minimally, "one human race per galaxy" and millions of "human races" in the universe. He cautions that this is not a certainty, and urges development of a "theology for these unknown worlds" which is open to the possibility of "their existence and presence."

Teilhard suggests, then, as theological starting points that if intelligence is not solely an Earth reality, "Christ must no longer be *constitutionally* restricted in his operation to a mere 'redemption' of our planet." As he continues in this vein, his thought becomes *anthropocentric* when he assumes that God became incarnate solely on Earth for all the cosmos. Extraterrestrial intelligent life, he assumed, would benefit from this earthly incarnation. He states that in order for "Christ's work of divinization" to extend cosmically, on each planet that has intelligent life God has revealed to "prophets and priests" that Christ's "redemptive Incarnation" has occurred. Because of cosmic unity, God need enter the cosmos only once "to permeate it in its entirety. . . . By taking a human nature, the Word was 'cosmified.'" Christ's death is "cosmic" in impact. Materiality itself has "*something of the Divine.*"

Teilhard's theological thinking in this regard is not only anthropocentric but *anthropomorphic*: he projects onto other worlds that their "hominized" life will need an Earth Christ's influence and "redemption." Remarkably, his idea that all intelligent life commits egregious offenses against God and harms itself is very like atheist scientist Stephen Hawking's assumption that evolved intelligent life inevitably tends toward self-destruction. A certain "cosmic anthropomorphism" is present in both Hawking and Teilhard, but emerges independently from their contrasting philosophical bases.

Teilhard had a slightly different take on the issue of incarnation and redemption in a previously unpublished essay written decades earlier: "Fall, Redemption, and Geocentrism" (1920). He suggested at that time, in words similar to those above, that it seems likely that intelligent life exists elsewhere in the cosmos. He speculates that evil is "contaminating" all of them. In a sharp contrast with his "prophets and priests" proposal just cited, he states that "the idea of an earth chosen *arbitrarily* from countless others as the focus of Redemption is one that I cannot accept; and on the other hand the hypothesis of a special revelation, in some millions of centuries to come, teaching the inhabitants of the system of Andromeda that the Word was incarnate on earth, is just ridiculous. All that I can entertain is the possibility of a multi-aspect Redemption which would be realized, as one and the same Redemption, on all the stars." Teilhard's thoughtful theological speculation about ETI is contrary to Protestant evangelicals' assertions that ETI must be baptized on Contact to "save their souls."

In Teilhard's thought, the cosmos is gradually proceeding from an Alpha Point (a divine act of creation that is the origin of the cosmos) to an Omega Point, where everything merges in a cosmic Christ. He is committed to Christian theology and he conservatively considers and interprets biblical texts that are its foundation (he generally has an almost literal interpretation of verses in the Christian Scriptures/New Testament that refer to an "original sin," divine incarnation in Jesus, divine redemption of a "fallen" Earth and cosmos, and a future emergence of Christ at the Omega Point). His ideas here are considerably inspired and influenced by the last chapter of the last book in the Christian Bible, Revelation. Jesus, who is seated on the throne of the new Jerusalem that has descended from the heavens to Earth, says to the seer, John, "I am the Alpha and the Omega, the first and the last, the beginning and the end" (22:12). For Teilhard as for John, all creation is simultaneously returning to and arriving at its source, which is both its origin and ultimate place, and final state of being. Then, all will be transformed and permeated by divine Being, expressed and presented biblically in the Cosmic Christ. Here, too, Teilhard's vision is a mix of conservative, literal biblical interpretation interspersed with his perspective that all creation ultimately will be united integrally and fully with its Creator, who already permeates all that exists. It is a teleological understanding in which ultimate divine intention and a divine attraction to relationship—not an absolute design—draw creation to its *telos*, its ultimate end.

Peacocke: Cosmic Convergence of Christic Concepts

Arthur Peacocke reflects on possible theological transference of the meaning and implications of divine incarnation, as expressed on Earth in the person of Jesus, to other cosmos contexts. In *Evolution: The Disguised Friend of Faith?*, Peacocke notes the possibility that ETI exist on other worlds as he reflects on a possible cosmic meaning or interpretation of divine incarnation: "If there is life on other planets, *as is at least possible*, what does this imply for the uniqueness of Jesus as Redeemer, Lord, Savior, and *Logos* incarnate?" (emphasis added). He does not respond to his own query; he suggests the issue as one that requires serious Christian consideration. However, in an earlier part of this book, Peacocke strives to present an understanding of incarnation that resonates with religious belief and biological data and can, in his estimation, provide an "inference to the best explanation" (IBE) that builds on the bases of theology and science. His resolution of an ideological conflict on Earth between biology and theology in regard to this issue might well, in this writer's opinion, serve to answer his question about incarnation as a doctrine applicable or at least acceptable in exoEarth contexts, for ETI beings. In these considerations, while Peacocke does not answer his question about what humans' perception of the uniqueness of Jesus as "Redeemer, Lord, and Savior" might mean elsewhere in the cosmos, the last phrase of his question is addressed: the uniqueness of Jesus as "*Logos* incarnate." The response, as will be elaborated below, is that Jesus is not "unique" as *Logos*-in-creation, except among humans (as far as is known or believed in Christian theology); the divine *Logos* will be incarnate among *all* intelligent life, in ways that God deems necessary and helpful, and to the extent to which intelligent beings are open to them, in the diverse form(s) of intelligent biota throughout the cosmos, whatever they might be.

Peacocke states that the understanding of "information input" in the field of communication theory provides a possible explanation of how Jesus could be fully human and fully express divinity. He cites John Bowker's proposals, in his book *The Religious Imagination and the Sense of God*, about the meaning of incarnation. For Bowker, Jesus can be regarded as a "wholly God-informed person, who retrieved the theistic inputs coded in the chemistry of brain processes for the scan of every situation, and for every utterance, verbal and nonverbal." This means Jesus embodied God "in the only way in which it could possibly have occurred" for a human being. What made Jesus particularly significant was the "stability and consistency with which his own life-construction was God-informed." On this basis, one could understand Jesus to be "a wholly human figure, without loss or compromise," and simultaneously "at exactly the same moment" a "wholly real presence of God." The divine "nature," whatever it is and to the extent possible, "can be mediated to and through the process of the life-construction in the human case, through the process of brain behavior by which any human being becomes an informed subject—but in this case, perhaps even uniquely, a wholly God-informed subject."

Peacocke, building on Bowker, suggests that Jesus as a human being responded entirely and directly to God's invitation to listen to God's teaching and guidance—an invitation offered to all people. In doing so, Jesus' words and actions always expressed the will of God: because of Jesus' openness of consciousness and response in conduct, God was fully revealed in and active through him. In this sense, he was fully divine, since God's will was what he chose to will, and divinity became expressed through his humanity. He was fully human as a biotic being, and fully divine to the extent to which every human who similarly responds to God is capable. Since no others have done so on Earth, Jesus is unique among humans and stands out as a God-revealing *Logos*-incarnate person.

The Bowker-Peacocke speculation about the Jesus-Spirit unity in one representative TI might be a kind of divine presence available to all intelligent beings—TI and ETI—who are similarly open to being informed and guided by the Spirit, and to expressing that Spirit in all they say and do. In terms of humanity, Peacocke states in *All That Is* that Jesus is the "distinctive manifestation of a possibility always inherently there for human beings by virtue of their potential nature. This makes what he *was* to what we *might be*. . . . What we have affirmed about Jesus is not, in principle, impossible for all humanity." Peacocke's ideas about Jesus', and potentially all humans', reception and revelation of the Spirit might be extended to be considered possible for selected members of every ETI species.

The Spirit's incarnation among ETI, in whatever physical form(s) ETI evolves throughout the cosmos, would be a similar activated and embodied expression of divine information, and thus of divine being. No "prophets and priests" would be required among ETI to intuit what God had done on Earth (as Teilhard suggested): each would have their own species' spiritual leaders whose teaching and lives would reveal and present (or would have revealed and presented in the past) divine presence and instruction. Peacocke himself, in his comments on the self-expression of the eternal creative

Word (*Logos*) who became incarnate in Jesus, states that while Christians perceive the uniqueness of Jesus, "God's Word, God's Self-expression, could also be 'manifest historically'" in "other religions and cultures." Peacocke's understanding here need not be confined solely to its Earth context. The Word that he says might be self-expressed among other Earth peoples, might be expressed, too, among ETI throughout the cosmos.

In at least two significant ways, Peacocke's proposal apparently contradicts basic Christian beliefs: that God became incarnate solely in Jesus in a "hypostatic union" of divine being-human being, and that this type of divine-human/divine-intelligent being union took place only in one intelligent being—a human on Earth—and nowhere else in cosmic creation. However, the traditional Christian doctrine about the wholly God-wholly human hypostatic union could be compatible with Peaocke's proposal: Jesus accepted and exemplified the divine-human bond (as Peacocke elaborates) and also, in some inexplicable way, was interrelated directly with and expressed the Spirit immanent in him.

Peacocke's preoccupation with and insistence on science-theology integration that does not violate the inherent integrity of either need not be accepted in its entirety by traditional or even progressive Christians. Many of them share his acceptance of and enthusiasm for scientific data, and his understanding that, since creation was initiated by a Creating Spirit, there cannot be a contradiction between what science discovers *in* creation and what theology teaches *about* creation, since both fields of inquiry explore aspects of divine creativity. However, many Christians are not willing to alter their beliefs about a "hypostatic union"—the manner in which Jesus was both God and a human being—or that God would not break the basic laws of God's creation, as discovered by science, in order to intervene in some way to affect particular events, such as when people ask for divine intervention in situations of extreme stress or on occasions of life-threatening illness. For Peacocke, such intervention would constitute a contradiction because in such an event present divine action would supersede initial divine action; Peacocke states that God's act of creation cannot be interfered with by God's interventional acts in creation.

Most Christians believe, too, that Jesus had healing powers that he compassionately used when he walked among his people, and that the Spirit of the universe has and will use divine healing power or another form of helpful intervention because of God's compassion. Indian peoples have a similar understanding, and native spiritual leaders and healers are known to have healing powers, or to be channels for the Creator Spirit's healing power—what some traditional healers call the "power of the universe." While Peacocke does not think or believe along these lines, his ideas might be extended to include Christian beliefs along these lines, as follows.

Peacocke, the priest-scientist concerned about the compatibility of theology and science as discovered by Earth-originating and -situated human scientists, does not discuss how the basic laws operative in the universe might not be the only characteristics of creation. As noted earlier (and as Peacocke accepts), God is a loving Spirit who is creative and free. God imparts to creation, at its origin, the freedom to be creative—autopoeisis—and God is immanent in creation as it unfolds—present through pan*en*theism. Peacocke suggests, along these lines, that the Creator is Being and Becoming, immanent and transcendent. The essential element of the Creator is *love*. Since creation emerges from the Creator who is love it, too, is permeated by love, an inherent characteristic. Further, since love is primary, prioritized over all other aspects of the cosmos, and since God is love, then the Spirit might creatively prioritize love in particular times and places—surely *kenosis* would be such an example—to provide a particular involvement in a situation or to meet an individual or communal need. God would *not* be violating God's own laws, since love is one of those "laws"; rather, God prioritizes love over the matter-energy aspects of the universe discernible to science. Moreover, God can intervene in this matter not by violating laws governing matter-energy-time-space. God's knowledge of the complexity and intricacy of what is operative now and what will unfold in the cosmos enables God to anticipate, appropriate, and apply currently unknown and unused possibilities through creatively reorganizing scientist-discovered laws and chronological sequences at particular moments. God's freedom to love is not eliminated by *kenosis*, the divine self-emptying at the origin of creation that enabled autopoeisis to happen; an overriding and continuing expression of divine love might be embodied or expressed in creation after the cosmic singularity. God can choose to prioritize love to benefit all or part of creation in particular circumstances. This exercise of the prioritization of love—a part of all-that-is—over other aspects of the cosmos would be a *precall* of the future rather than a *recall* of the past or rearrangement of what is present.

Christian theology describes *anamnesis*, a making present of a past moment and event (for example, in Catholicism, when the priest at Mass celebrates the Last Supper, Jesus and his words and actions become present at the moment of the priest's consecration of the bread and wine). The Spirit's intervention to benefit a person or community would be an *anticipatory anamnesis*: again, a *precall* of what is to come, making it present in this place and time. God's creative love at a particular present cosmic moment is prioritized over God's initial and ongoing creative love, initially expressed in the singular moment of initial creation of a dynamic cosmos and continuing to be present during autopoeisis. Analogously, loving parents give their children responsible freedom, but at times restrict it or intervene with loving compassion and concern when children ask for help (and sometimes when they don't) at different times in their life—but then ease away, respecting and encouraging their children's autonomy, creativity, and contextual decision-making as they exercise responsible freedom by making responsible choices.

Consideration of the ideas just presented might make more acceptable, for the traditional Christian, many of Peacocke's ideas, including his key thought that there is no contradiction between science and theology, and that God would not intervene in history—in time and space—in order to respond to a situation of compelling need, by breaking laws God created for the universe. Regarding love as one of those laws, and recognizing divine power to utilize future unfolding of the matter-energy interrelationship to resolve present circumstances, would enable people to accept Peacocke's fundamental presupposition in this case.

Omega Point: Cosmic Christ, Cosmic Logos

What might be drawn from the ideas of Teilhard and Peacocke is that as each intelligent species accepts and follows the Spirit-Presence invitation and guidance it evolves socially, psychologically, and spiritually. Only one Spirit is present in the cosmos, and in diverse worlds of the cosmos, in different forms, it is a loving, teaching, guiding power who becomes part of each world in ways and in times as prompted by the Spirit's love. The process of worlds' spiritual evolution is not dependent on the spiritual or material entry of divine Spirit amid solely members of one species on one planet in the entire universe, a Presence somehow known to have occurred once and for all in some way by spiritual seers throughout the cosmos. Each and every world, at times and in diverse cultures and places in its history and according to its needs, has spiritual seers who particularly embody and present the immanent-transcendent, Being-Becoming, loving and creating

Spirit, Science, and Space

Spirit's Presence. The end result of cosmic evolutions is that the Spirit in each place—the eternal *Logos*-Word—is drawing all to a common place at an Omega Point—perhaps a point for each cosmos or, even as space expands and discrete cosmic segments might not any longer be individually and mutually accessible universally, a place and consciousness to which every cosmos is drawn. (Interdimensional travel, if it becomes a reality, could mean that cosmic separation would not occur entirely, despite increasingly vast spatial distances among galaxies or star systems.)

The "Cosmic Christ" envisioned and expressed by Teilhard as the Omega, the end toward which the cosmos is headed, might be extended universally: a Cosmic Spirit—as creating *Logos*—has been present to and understood by diverse species. This cosmic Spirit has had distinct manifestations in diverse worlds, has been interpreted in culture-specific ways in each world, and through intelligent beings' social and spiritual evolution has come to be understood and accepted on each world as globally species-specific and not regionally or intellectually culture-specific. The *Logos* presence is not institutionalized in a religion or theologized in academia, but is related in community to all members of a species. The "Cosmic Christ" is a local, terrestrial concept and understanding, seen through human eyes and understood by human culture, specifically among Christians. The Cosmic Spirit is a cosmic concept and reality—more extensive, accessible, and universal. The Spirit, as creating and immanent *Logos*-Word, draws all things throughout the cosmos to a fuller immersion into the spiritual dimensions of reality, in ways specific to each culture's needs—perhaps by being embodied in intelligent species' physical forms, in chosen members of individual species, in a transformative way. This would be an extension of Maximus' teaching that there is a *Logos-logoi* dialogic relationship in all creation: there are elements of the creating *Logos* in all the created *logoi*, which include all abiotic and biotic being.

Cosmic integral being is interrelated in and with the Being-Becoming immanent Spirit who is in dialogic engagement with all-that-is, who becomes part of divine creation in a special way at sacred moments in select intelligent beings. This understanding gives new meaning to verses in the early Christian hymn that opens the Gospel according to John:

> In the beginning was the Word, and the Word was with God, and the Word was God. He was in the beginning with God. All things came into being through him, and without him not one thing came into being. What has come into being in him was life, and

> the life was the light of all people.... He was in the world, and the world came into being through him; yet the world did not know him. He came to what was his own, and his own people did not accept him. But to all who received him, who believed in his name, he gave power to become children of God, who were born, not of blood or of the will of the flesh or of the will of man, but of God. And the Word became flesh and lived among us, and we have seen his glory, the glory as of a father's only son, full of grace and truth. (John 1:1–14, abridged)

The meaning of this passage, if it were to be applied cosmically simply by changing "world" to "worlds," is extraordinary and could stimulate on Earth and on other worlds a greater appreciation for and love of the Spirit, while expanding intelligent beings' consciousness into the cosmos. Arthur Peacocke's commentary on the verses, in *All That Is*, could be extended on a cosmic scale. Peacocke observes that "the Johannine concept of the *Logos*, the Word of God, may be taken to emphasize God's creative patterning of the world and hence as God's self-expression *in* the world"—and, it might be added, "and *in* all the *worlds* of the cosmos."

Complementary understandings of how God might be ever more present in those open to a heightened experience of divine immanence might have been expressed, too, by Paul in the Christian Bible. In his letters, Paul declares that he is an apostle committed to doing the work of the Spirit. He says that "it is no longer I who live, but Christ lives in me" (Gal 2:20). In another letter, Paul states that "those who are spiritual discern all things, and they are themselves subject to no one else's scrutiny. 'For who has known the mind of the Lord so as to instruct him?' But we have the mind of Christ" (1 Cor 2:16). Perhaps Paul, along lines suggested by Bowker-Peacocke, was open to and accepted God's guidance to an ever-increasing degree (beginning when he converted from a persecutor of Christians to a committed follower of Christ), but not fully or in a "hypostatic union" relationship with God such as Jesus had according to Christian doctrine. (I'm not trying to "proof text" or do *eisegesis* by presenting these verses in order to back me up. Rather, I'm suggesting that the verses *might* provide insight into what Paul experiences and is writing.) Paul's thoughts are related to the Bowker-Peacocke idea that Jesus accepted to an extraordinary degree the union of his will and God's will, and that this degree of acceptance of the divine will might be available to others. Christian theologian Sallie McFague expresses this idea in *Blessed Are the Consumers: Climate Change and the Practice of Restraint*. She says that "Jesus, as the face of God, tells us that who Jesus 'is' manifests itself in friendship with God—his will is totally open to God's will, wanting to do only what God wishes, no matter the cost to the self, even death on a cross.... The mysterious doctrine of the hypostatic union can be interpreted as oneness of intentionality between God and Jesus, in which Jesus empties his own will in order to be filled with the will of God."

The atheist, not accepting a Cosmic Christ or that a universal Spirit becomes present in some way in every intelligent species, might understand instead that there is a permeating dynamic force or energy that is operative in the cosmos. In this view, the immanent power understood by theists to be the Spirit of the universe might be viewed by secular humanists as a form of yet-to-be-discovered energy.

Francis and Funes

In his "Canticle of All Creatures," Francis of Assisi (1182–1226) calls abiotic beings "brother" or "sister": Brother Sun, Sister Moon and stars; Brother Wind and air and cloud, and Sister Water; Brother Fire, and "our Sister, Mother Earth." These written verses are complemented by the words of the Italian romantic ballad to whose melody Francis sang his canticle during his journeys. The anonymous composer's original verses celebrated the beauty of biota. In the canticle Francis sang then, in his words and in the original words his hearers would know because of the melody he used, all creatures, biotic and abiotic, are related familially each to the other. The Earth- and biota-related spirituality of Francis in the thirteenth century is complemented by the centuries- or millennia-old Earth- and biota-related spiritual understandings of the Lakota, whose tradition teaches that "we are all related." Both describe familial and communal understandings of the ways in which humans are to relate to each other and to all creatures.

José Gabriel Funes' *L'Osservatore Romano* interview, cited earlier, could be fruitfully recalled here. As noted, in the spirit of Saint Francis he suggests how we might approach, appreciate, and accept arriving ETI. Francis' song and Funes' statement together take on enhanced meaning if the *Logos*-Word has indeed become uniquely revealed in exoEarth milieus. We are all one family and community, related through the *Logos* whom we all bear within. The Hindu greeting *namasté* expresses this well: "The Spirit in me greets the Spirit in you. The Spirit in you greets the Spirit in me. The Spirit in you greets the spirit in me. The Spirit in me greets the spirit in you. The spirit in me greets the spirit in you. The spirit in you greets the spirit in me." We are all related spiritually, on multiple levels and in multiple dimensions of reality, as well as materially.

Spirit and space are integrated in this way of thinking, and in complementary insights from Indian cultures throughout the U.S. Everything in the universe has emerged from and exists in the Creator Spirit; TI and ETI exist in creation and are related to the whole community of living beings; all are "people" in an interrelated holistic cosmos.

Space has the potential to be a significant cosmic integrating factor in the relationship of religion and science, much as ecology has been in their relationship to each other on Earth. As secular humanist scientists and theistic scientists, and nonscientist atheists and theists came to realize, a coalition of faith communities and environmental communities, working together despite their metaphysical differences, could make great strides toward saving not only life on Earth, but abiotic planet Earth. Currently, integrated efforts to stop global climate change, which is causing regional and local environmental havoc and ecological destruction and displacement, are being undertaken by scientists, theists of all professions, and environmental and social community activists and leaders. A similar integration of efforts will enable not only better cooperation in cosmos explorations, but better Contact and communication with ETI—particularly if both TI and ETI are open to whatever, if any, spiritual insights the other has. Should this openness and communication be mutually embraced, then to an ever-greater extent Spirit, science, and space will be integrated by intelligent beings in the totality of consciousness of the integral being of the cosmos.

CONCLUSION

We Aliens Are Going! We Aliens Are Coming!

OUR INTELLECTUAL VOYAGE ON our thought experiment is coming to an end—at least in terms of *Encountering ETI*, but not necessarily in regard to encountering ETI. We do not know how soon—or even if—ETI will initiate sustained Contact with us. Only when that occurs will people be vindicated who have reported—and those who have seen but feared to report—that they have witnessed UFOs that are inexplicable as ordinary meteorological phenomena or human artifacts. Then, too, will U.S. government data collected for the past seventy years or so finally be released (we hope), and honest skeptics finally will be convinced. They would not need at that point to have physical evidence of ETI that is "beyond a reasonable doubt" since a sustained Contact should erase any doubts—except, of course, for the types of people who still deny that Neil Armstrong walked on the moon: they claim that the video scenes of the lunar landing that were broadcast live were fabricated.

Aliens Here and There

To what extent will we be "aliens" on and to Earth when we voyage to the stars? Will our consciousness and conduct still alienate us from our home planet and from ETI who are unlike us (whoever "we" are who course the cosmos)? We will then, as aliens not only on other celestial bodies but on our Earth home, always be "out of place" wherever we are and wherever we venture. However, if we can undergo a significant change in our understanding and see ourselves as part of "nature" (as planet and as biotic community) in our terrestrial home here, then we will be able to see ourselves as part of "nature" elsewhere, far out into the universe. We will understand, to the extent possible, that we are part of integral being, of the cosmic whole. If we remain aliens on Earth, as evidenced by our conduct in our home environment and toward our kin, the flora and fauna and all else living that together comprise the biotic community, we will be unable to be more than aliens wherever we go. Extraterrestrial intelligent beings will react in fear ("Aliens!") at our approach. However, if we are at home on Earth we will be at home in the heavens, and by our conduct ETI will recognize and respond to us not as alien "Others" but as companion intelligent beings who have not left their home alienated, and do not journey as perpetual "aliens." We will be greeted with xenophilia, not rejected because of xenophobia. We will be welcomed as stimulating and perhaps even exciting visitors.

We have the potential to be, for others we encounter while exploring in space or after landing on other worlds we find, one of the two alien stereotypes of our sci-fi films. We could be benevolent beings who on Contact and thereafter will enhance our own and others' growth into being cosmic citizens; or, we could be malevolent beings who range the cosmos to attack and even annihilate or otherwise subordinate to us other ETI, or do whatever else seems necessary to acquire native residents' natural goods. We might use an updated Discovery Doctrine and an Adam Smith ideology in our exploration, or an adaptable-to-context Outer Space Treaty and Thomas Paine philosophy. We should ask ourselves, before we enter our Star Age, the question, which type of alien do we want to most closely approximate in our space travels? If we aspire to be benevolent aliens in our travels and on Contact, what changes in consciousness and conduct must we begin to make now, before our departure to the stars?

Tradition, Transition, and Transformation

As has happened before in human history, paradigm shifts are in the offing. As noted before in this book, a paradigm shift can occur for humankind not only on Contact, but even during consideration of the possibility of Contact. Our response to the Roswell and Hudson River Valley narratives, even when we ponder them solely "as if" they occurred without weighing in to take sides in the debate over their historicity, can profoundly affect our psyche for better or for worse. It can catalyze us to enter into a cocoon where profound insights and intents can be formed in our chrysalis state while we ponder with our intellect and in our spirit our current ethical,

Conclusion

ecological, economic, and ecclesial understandings of the cosmic dimensions of reality and life on Earth, such that we are slowly transformed. In this case, we will emerge with wings to soar into space and expand our sense of being citizens of the cosmos. Or, we can be traumatized just by entering a cocoon laden with a new consciousness and data, which would cause us to strive to remain encased (which would inevitably lead to death) or try to go back to our pre-chrysalis state (which is physically, spiritually, and intellectually impossible) rather than experience transition as we are being formed and transformed. In this case, we will be doomed: either we will be unable to emerge from our cocoon because we have not changed sufficiently to do so, or we will emerge deformed and deteriorating because we stopped our transformation. We can freeze tradition, refusing to recognize that every tradition had a starting point, and new traditions await development from and upon the foundation of the old. We can, then, begin to develop new traditions that incorporate the old in an integrative transitional state. This will lead to our transformation from being earthbound (tied to old understandings of Earth) to being Earth-integrated (interrelated with our place and with our interdependent relationships with Earth and with biotic being—in general and as particularized in our own species and its diverse communities). We can accept *intra*generational and *inter*generational responsibility. We can excitedly participate in a dialogic relationship between present and future space and time: between celestial contexts here and elsewhere, and between present and future temporality. In space and time each can stimulate the other, to their mutual benefit.

Cosmosocioecological Praxis Ethics

Say what? You probably blinked at least twice when you saw that heading. Rest assured: it'll be broken down into its component parts to be very accessible—and acceptable? (I refer readers seeking a more in-depth discussion on this and other topics to consult *Cosmic Commons*. It even has footnotes, an index, and a bibliography!)

Socioecological ethics weaves together principles and practices of *social ethics* (promoting social justice in and among human communities) and *ecological ethics* (promoting ecological wellbeing among humans, between humans and other life in the biotic community, between humankind and Earth, and among humans, other biota, and Earth) in an integrated theistic perspective. Socioecological ethics is a Spirit-related *creatiocentric* (creation-centered, which relates Creator and creation; it is not theocentric, anthropocentric, androcentric, egocentric, gaiacentric, geocentric, or biocentric), reflective integration of ethical theory and method that stimulates human striving for social justice (right conduct in community) and humans' sustaining commitment to ecological wellbeing (right conduct in relation to abiotic and biotic creation). It includes consciousness of the immanent-transcendent Creator in and with whom all being coexists in cosmic, Spirit-immanented integral being. In a humanist perspective, *socioecological ethics* is a transcending values-related terracentric method that interrelates social justice (right conduct in community) and commitment to ecological wellbeing (right conduct in relation to Earth's abiota and biota) with consciousness of the human interrelationship with all being. Both perspectives express *Mitakuye oyasin*—"we are all related"—and *namasté*—our Spirit and spirits are in relation.

The *praxis ethics* process ("methodology") is the contextualization of principles in practice. It adapts to and from context but retains its core values while willing to reasonably negotiate, on site, the secondary considerations that it brought to context based on prior experiences; its priority is not to implement what it brings to context, as in the field of applied ethics, but to bring to and learn from context what consciousness and conduct might serve best to effect socioecological justice and wellbeing. In context, it promotes the commonweal in the social and bioregional commons. Praxis ethics seeks *common ground*, in three senses: first, it is done "on the ground," rather than through abstract philosophical or theological speculation and formulation, second, it is done on the shared common ground of all life—Earth as home and habitat (and exoEarth worlds as potential homes and habitat); and third, it seeks to inspire theoretical and practical common ground: a shared understanding of historical context, shared principles, a shared vision, and shared social projects that incorporate people's complementary talents and insights, and respect their degrees of engagement in common collaborative efforts to promote socioecological ethics and eco-justice. It is the reflective integration of a dynamic, dialogical, and dialectical continuum of social context and social theory, intended to promote the commonweal in the social and bioregional commons, in Earth and exoEarth contexts. Praxis ethics includes integration by *inclusion* or by *exclusion*. *In*clusion occurs when new data, new perceptions of data, or a new social setting which inform ethics are accepted and incorporated. *Ex*clusion occurs when the new data, perceptions, or context that might at first sight appear able to inform (in thought or through events) the dialogic process are rejected or partially rejected as unrealistic or inappropriate for this milieu, or contrary to core values and principles. Praxis ethics is an on-the-ground exposition and enactment of socioecological ethics in specific contexts.

Cosmosocioecological ethics extends and enhances socioecological ethics to embrace the seen and unseen, known and unknown dynamic universe, a cosmic unfolding of creation (and/or autopoeitic generation) that is continuously *in statu viae* (a state of being-becoming). Humankind and all intelligent life is responsible for the wellbeing of those parts of the cosmos with which they become particularly interrelated when they pass through as explorers, or when they become present as communities of colonizers and conscientious developers of natural goods.

Commonweal: Common Wellbeing on Common Ground

Socioecological praxis ethics promotes in diverse contexts the Earth *commonweal*, other worlds' *commonweal*, and a cosmic *commonweal*. TI by itself or in collaboration with ETI prioritizes and works toward a common wellbeing, which includes the wellbeing of all individuals, rather than solely individual wellbeing. Individuals who do not recognize their community relationship and community needs might seek to promote their selfishness or at least their self-interest, neither of which includes sufficient consideration or commitment to include the wellbeing of communities as a social whole.

On Earth's or other worlds' common ground, TI and ETI would have as a core principle "Love your neighbor" to the greatest extent. "Neighbor" is known to mean all biotic and abiotic beings near and far. This perspective in practice means to strive to promote the "greatest good for all" rather than the "greatest good for the greatest number." This would result in the wellbeing of all, to the greatest extent possible.

In practice, operative positive principles would be "Do *for* others as you would have others do for you"; "Do *to* others as you would have others do to you"; "Do *with* others as you would have others do with you"; and operative negative principles would include "Do *not* for others what you would not want others to do for you"; "Do *not* do to others what you would not want others to do to you"; and "Do *not* do with others what you would not want others to do with you." These principles are especially important for inter-world encounters. The extent to which they are observed will indicate the extent to which TI and ETI, as individual or interacting species, are benevolent aliens or malevolent aliens.

Cosmosocioecological Praxis Ethics Dialogically Engaged

While Earth and Earth's communities are being transformed, humankind will be venturing into space to explore the moon, planets, and other celestial bodies. Cambridge scientist Stephen Hawking advocates space exploration and colonies because Earth is in peril. However, if attitudes are not changed on Earth, space explorers will export economic, ecological, ethical, ecclesial and other problems to distant places. The quest for natural goods to support the human adventure and to enhance life on the home planet requires a new consciousness and new ethical commitments in order that people might respect and relate well not only to existing natural ecologies in new contexts, but also to the evolving life—intelligent or less developed—that they encounter.

TI care for creation in terrestrial and extraterrestrial settings would facilitate engagement with ETI, and development of mutually beneficial policies and procedures. The practice of cosmosocioecological ethics described above would be helpful for this to happen. A four-step process of contextualizing cosmosocioecological ethics might be undertaken sequentially or concurrently, depending on planetary or interplanetary contexts and needs. In all of these steps, ordinarily a policy of noninterference would be operative as TI or ETI aliens observe from afar developing intelligent life on other worlds. However, perhaps on rare occasions it might seem appropriate to prioritize the primary principle of love of neighbor in a particular planetary context by suggesting at least symbolically that the intelligent beings are not on the right track for promoting planetary community wellbeing. (Reports by Lieutenant Salas in Montana and Colonel Plantonev in Russia that ETI had shut down ICBMs might indicate that ETI was sending a message to both superpowers.)

In the first step, *social analysis*, efforts would be made to discern the ideological bases—including, on Earth, the ongoing remnants of Discovery Doctrine—that have promoted humans' degradation of Earth and disparagement of Earth peoples. Evaluating these bases in human consciousness would lead to discussion of humans' conduct by assessing practices that have promoted the socially and ecologically malevolent ideology. An honest evaluation of the expressed or hidden reasons for space exploration would be made. This would stimulate consideration of possible harmful consequences, onto other planets, of Earth's dominant, current, rapacious ideology. In places where social structures already exist when humankind arrives, a similar social analysis would be made of present social inequities and biases among ETI, if any, including in the procurement of natural goods and the production and distribution of ETI goods that result from ETI labor; and, of ETI-caused or exacerbated ecological perils, if any, and the ideologies that foment them. The process of mutually engaged social analyses and the data it develops would lead to consideration of an alternative consciousness and conduct based on universally agreed upon values and ideals.

The second step, *social insight*, would be comprised, on Earth, of teachings derived from such sources as social scientific findings and U.N. documents such as the Universal Declaration of Human Rights, the Declaration on the Rights of Indigenous Peoples, the World Charter for Nature, and the Treaty on Principles Governing the Activities of States in the Exploration and Use of Space, Including the Moon and Other Celestial Bodies (the Outer Space Treaty). In extraterrestrial milieus, native intelligent beings' pre-Contact traditions, particularly as expressed in spiritual understandings and socioecological ethics, would be sought and included (an effort not undertaken in the provocative and prophetic film *Avatar*).

In the third step, *social vision*, the collaborative consciousness of human explorers, entrepreneurs, scientists, engineers, government officials, and Earth citizens will develop new Earth and exoEarth perspectives. Their communal vision will project their common hope for TI: that we will be new people in a new and just social order and a renewed Earth. Humans' beneficent social and ecological practices will have resulted from consideration and implementation of socioecological values and principles such that theory and practice have become congruent . . . and could be contextualized as starting points wherever TI might journey. People will be integrated in, interdependent among, and interrelated with other biota and abiotic Earth and exoEarth worlds. The vision might be expressed through a Cosmic Charter.

The fourth step, *social project*, would be to develop collaborative work that would gradually lead to the contextualization of the social vision previously formulated. In concrete contexts, the vision and the proposed conduct that it suggests could be altered as necessary to accommodate diverse celestial contexts, and adjusted as needed to facilitate mutually productive encounters with ETI who have their own distinctive consciousness and conduct. TI might realistically hope that in extraterrestrial contexts, a terrestrially formulated cosmosocioecological ethics will be able to evolve contextually, without essential compromise, in response to encounters with ETI.

Encounters between TI and ETI seem inevitable, given the age, extension, continuing evolution, and ongoing complexification of the cosmos. Intelligent beings have most probably been gradually emerging and evolving throughout the universe, sequentially on particular worlds and simultaneously in distant star systems. Such emergent events might occur at every cosmic instant, and therefore be in various stages of development throughout the universe at any moment of time as measured by TI on Earth, or by ETI on their worlds.

In order for humankind to promote and be part of planetary, interplanetary, and universal wellbeing, humans must gradually develop a sense of their place in Earth and cosmic contexts. Contact with ETI, if and when it occurs, will require each participant to consider carefully the implications of the other's principles and practices and, in the interests of peace to be generated through interspecies' and inter-celestial bodies' cooperation, to work to resolve, in context, such differences in perspective and practice as have emerged in distinct socioecological cosmic milieus. When both TI and ETI have a cosmic consciousness and a practical foundation in terrestrial-extraterrestrial cosmosocioecological praxis ethics, their perspectives will be informed, enhanced, and stimulated as they explore cosmic realities, elaborate common responsibilities, and establish parameters for realistic concrete projects. The result would be responsible scientific research and outreach, respectful terrestrial-extraterrestrial engagement, and spiritual receptivity to the implications and impacts of Contact. In our pre-ETI life we could at least theorize, as TI, what is *our* cosmic place, at least at *this* moment in time, and update our thinking when we adjust to ETI encounters in succeeding times and places.

Peoples of diverse theistic faith traditions, when mindful of divine Presence, can care for creation in its terrestrial and extraterrestrial ecosystemic manifestations, and respect diverse biotic communities—evolving sociobiological extraterrestrial life and sociocultural extraterrestrial intelligence—wherever humans' theoretical considerations and concrete explorations take them. Peoples of diverse humanist faith traditions can do likewise, based on their own ideals and visions for the future of humankind. Both types of traditions, with complementary and even congruent dreams and hopes, together can concretize in Earth and exoEarth socioecological settings, through collaborative projects, the *utopias* they share in common, the ideal places they envision that have yet to be developed and attained.

Eve of World(s) Destruction or Dawn of Cosmic Community?

As we look at the socioecological state of Earth at this time in human history, we see species being extincted by humankind and Earth being devastated by humankind. We have glaring economic disparities, and wars and rumors of wars. Singer Barry Maguire expressed well current human conduct in his song "Eve of Destruction" (1965), one verse of which notes the beginning of humankind's ventures into space: "Ah, you may leave here for four days in space, but when you return it's the same old place. . . . Hate your next door neighbor, but don't forget to say grace, and you tell me, over and over and over and over again my friend, Ah, you don't believe we're on the eve of destruction." Will we continue to hate our next-door—or distant—neighbor, even as we travel in the cosmos and when we return to Earth? Are we on the eve of our own world's and other worlds' destruction? The answer, lamentably, is yes—if we continue our operative consciousness and conduct on Earth, and export it into the cosmos.

We must consider seriously whether or not Stephen Hawking is right in saying that inevitably we will destroy ourselves and our planet. Perhaps, instead of being correct about the future, Hawking is prophetic in the sense of the seers of ancient Israel: we will destroy ourselves *if* we do not change our ways and convert from being native malevolent TI to native benevolent TI—with a corresponding conduct change in near and outer space. Haudenosaunee elder Oren Lyons, after discussing the centuries-old prophecies about ecological devastation described by Seneca elder Handsome Lake, reflected on a coming time when the prophecies possibly will be fulfilled, which he saw as inevitable. Lyons taught that each generation must take the responsibility to ensure that environmental destruction does not occur in their time. If all of us humans, we who live today and our progeny in generations to come, strive to avert socioecological catastrophe in our time, then not only will we help to postpone such an event (if it is indeed inevitable), but we will continually better our biotic and abiotic relationships. Earth and all life will benefit if we do so. We could be on the dawn of a real Earth community, and then of cosmic community. If we think and act responsibly, the dawn will gradually usher in a new day.

In our new consciousness, we can extend John Donne's classic poetic statement that "no man is an island," and declare that "no species is an island" and "no planet is an island." In integral cosmic being, we are interrelated and bonded together. We will become the aliens that we hope will come to Earth. Then we can say: we go in peace from our world into the cosmos. We come in peace on other worlds throughout the cosmos.

SELECT BIBLIOGRAPHY

Sources for *Encountering ETI* material and for further reading; in lieu of text footnotes, select ranges of pages from some books are indicated.

Archbishop Demetrios. "Love and Care for Strangers: *Philoxenia* in Christian Tradition." *BTI Magazine* 12.1 (2012) 4–7.
Ayala, Francisco J. *Darwin and Intelligent Design*. Minneapolis: Fortress, 2006.
———. *Darwin's Gift to Science and Religion*. Washington, DC: Joseph Henry, 2007.
BBC. "Churchill Ordered UFO Cover-Up, National Archives Show." August 5, 2010. http://www.bbc.co.uk/news/uk-10853905.
———. "Files Reveal 'Rendlesham Incident' Papers Missing." March 2, 2011. http://www.bbc.co.uk/news/uk-12613690.
———. "Ministry of Defence Files on UFO Sightings Released." August 11, 2011. http://www.bbc.co.uk/news/uk-14486678.
———. "MOD Releases UFO Files." May 14, 2008. http://news.bbc.co.uk/2/hi/uk_news/740023.
Berliner, Don, and Stanton T. Friedman. *Crash at Corona: The U.S. Military Retrieval and Cover-Up of a UFO*. New York: Paraview, 2004.
Bird, Joan. *Montana UFOs and Extraterrestrials: Amazing Stories of Documented Sightings and Encounters*. Helena, MT: Riverbend, 2012.
Bradbury, Ray. *The Martian Chronicles*. New York: Harper Perennial Modern Classics, 2011.
Butler, Smedley D. *War Is a Racket*. Port Townsend, WA: Feral House, 2003.
Carey, Thomas J., and Donald R. Schmitt. *Witness to Roswell—Unmasking the Government's Biggest Cover-Up*. Pompton Plains, NJ: New Page, 2009.
Churchill, Ward. *A Little Matter of Genocide: Holocaust and Denial in the Americas, 1492 to the Present*. San Francisco: City Lights, 1997.
Clarke, Ardy Sixkiller. *Encounters with Star People: Untold Stories of American Indians*. San Antonio: Anomalist, 2012.
Corso, Philip J., with William J. Birnes. *The Day After Roswell*. New York: Pocket, 2008.
Cox, George W. *Alien Species in North America and Hawaii: Impacts on Natural Ecosystems*. Washington, DC: Island, 1999.

Select Bibliography

Cronon, William. *Changes in the Land: Indians, Colonists, and the Ecology of New England.* New York: Hill and Wang, 1983.

Crowe, Michael J. *The Extraterrestrial Life Debate—Antiquity to 1915.* Edited by Michael J. Crowe. Notre Dame: University of Notre Dame Press, 2008.

Cuénot, Claude. *Teilhard de Chardin: A Biographical Study.* Translation from *Pierre Teilhard de Chardin: les grandes étapes de son evolution* by Vincent Colimore. Edited by René Hague (Librairie Plon, 1958). Baltimore: Helicon, 1965.

Darwin, Charles. *On the Origin of Species.* A Facsimile of the First Edition. Cambridge: Harvard University Press, 1964.

Dick, Steven J., ed. *Many Worlds: The New Universe, Extraterrestrial Life and the Theological Implications.* Philadelphia: Templeton Foundation, 2000.

Dowd, Michael. *Thank God for Evolution—How the Marriage of Science and Religion Will Transform Your Life and Our World.* New York: Plume, 2009.

Farrell, John. *The Day Without Yesterday—Lemaître, Einstein, and the Birth of Modern Cosmology.* New York: Thunder's Mouth, 2005.

Foner, Philip S., ed. *The Life and Major Writings of Thomas Paine.* Secaucus, NJ: Citadel, 1974.

Friedman, Stanton T. *Top Secret/MAJIC—Operation Majestic-12 and the United States Government's UFO Cover-Up.* Philadelphia: Da Capo, 2008.

Hardin, Garrett. "Living on a Lifeboat." *BioScience*, October 1974, 561–68.

———. "The Tragedy of the Commons." *Science*, December 13, 1968, 1243–48.

Hart, John. *Sacramental Commons: Christian Ecological Ethics.* Lanham, MD: Rowman & Littlefield, 2006.

———. "Cosmic Commons: Contact and Community." *Theology and Science* 8.4 (2010) 371–92.

———. *Cosmic Commons: Spirit, Science, and Space.* Eugene, OR: Cascade, 2013.

Haught, John F. "What if Extraterrestrials Exist?" In *Science and Faith: A New Introduction*, 163–76. Mahwah, NJ: Paulist, 2012.

Holder, Rodney D., and Simon Mitton. *Georges Lemaître: Life, Science, and Legacy.* Astrophysics and Space Science Library. London: Springer, 2013.

Hui, Sylvia. "Physicist Hawking Says Human Race Must Look to Outer Space for Survival." *Helena Independent Record*, June 14, 2006, 5A.

Hynek, J. Allen. *The Hynek UFO Report.* New York: Barnes & Noble, 1997.

———. *The UFO Experience: A Scientific Inquiry.* Chicago: Henry Regnery, 1972.

Hynek, J. Allen, et al. *Night Siege: The Hudson Valley UFO Sightings.* 2nd ed. St. Paul, MN: Llewellyn, 1998.

Kaminski, Bob. *Lying Wonders: Evil Encounters of a Close Kind.* Mulkiteo, WA: Wine Press, 2006.

Kean, Leslie. *UFOs: Generals, Pilots, and Government Officials Go on the Record.* New York: Three Rivers, 2010.

King, Ursula. *Spirit of Fire—The Life and Vision of Teilhard de Chardin.* Maryknoll, NY: Orbis, 1998.

Leake, Jonathan. "Don't Talk to Aliens, Warns Stephen Hawking." *The Sunday Times*, April 25, 2010. http://timesonline.co.uk/tol/news/science/space/article7107207.

Lukas, Mary, and Ellen Lukas. *Teilhard: The Man, the Priest, the Scientist.* Garden City, NY: Doubleday, 1977.

Lyons, Oren, and John Mohawk, et al., eds. *Exiled in the Land of the Free: Democracy, Indian Nations and the U.S. Constitution.* Santa Fe, NM: Clear Light, 1992.

Select Bibliography

Mack, John E. *Abduction: Human Encounters with Aliens*. New York: Scribner, 2007.
———. *Passport to the Cosmos—Human Transformation and Alien Encounters*. New York: Crown, 1999.
Mann, Charles C. *1493: Uncovering the New World Columbus Created*. New York: Knopf, 2011.
Marcel, Jesse, and Linda Marcel. *The Roswell Legacy: The Untold Story of the First Military Officer at the 1947 Crash Site*. Franklin Lakes, NJ: New Page, 2009.
Miller, Robert J. *Native America, Discovered and Conquered: Thomas Jefferson, Lewis and Clark, and Manifest Destiny*. Native America: Yesterday and Today. Westport, CT: Praeger, 2006.
Miller, Robert J., et al. *Discovering Indigenous Lands: The Doctrine of Discovery in the English Colonies*. Oxford: Oxford University Press, 2012.
Mondragón Cooperative Movement. http://www.mondragon-corporation.com/.
MSNBC. "Hawking: Aliens May Pose Risks to Earth." April 25, 2010. http://www.msnbc.com/id/36769422/ns/technology_and_science-space/.
Newcomb, Steven T. *Pagans in the Promised Land: Decoding the Doctrine of Christian Discovery*. Golden, CO: Fulcrum, 2008.
Paine, Thomas. *Agrarian Justice*. In *The Life and Major Writings of Thomas Paine*, edited by Philip S. Foner, 605–23. Secaucus, NJ: Citadel, 1974.
———. *The Rights of Man–Part Second*. In *The Life and Major Writings of Thomas Paine*, edited by Philip S. Foner, 345–458. Secaucus, NJ: Citadel, 1974.
Peacocke, Arthur. *All That Is—A Naturalistic Faith for the Twenty-First Century*. Edited by Philip Clayton. Minneapolis: Fortress, 2007.
———. *Creation and the World of Science*. Oxford: Clarendon, 1979.
———. *Evolution: The Disguised Friend of Faith?* London: Templeton Foundation, 2004.
———. *Paths from Science towards God: The End of All Our Exploring*. Oxford: Oneworld, 2001.
———. *Theology for a Scientific Age: Being and Becoming—Natural, Divine, and Human*. Minneapolis: Fortress, 1993.
Peters, Ted, and Martinez Hewlett. *Evolution from Creation to New Creation—Conflict, Conversation, and Convergence*. Nashville: Abingdon, 2003.
Salas, Robert L., and James Klotz. *Faded Giant*. BookSurge, 2005.
Smith, Adam. *The Wealth of Nations*. New York: Modern Library, 2000.
Stannard, David E. *American Holocaust: The Conquest of the New World*. New York: Oxford University Press, 1992.
Swords, Michael, and Robert Powell, eds. *UFOs and Government: A Historical Inquiry*. San Antonio: Anomalist, 2012.
Teilhard de Chardin, Pierre. *The Divine Milieu*. Rev. ed. New York: Harper & Row, 1968.
———. *The Human Phenomenon—A New Edition and Translation of* Le phénomène humain. Translated by Sarah Appleton-Weber. Eastbourne, UK: Sussex Academic, 2003.
———. *The Phenomenon of Man*. 2nd ed. New York: Harper & Row, 1965.
———. "A Sequel to the Problem of Human Origins: The Plurality of Inhabited Worlds." In *Christianity and Evolution*, translated by René Hague, 229–36. New York: Harcourt, 1974.
Tinker, George E. *American Indian Liberation: A Theology of Sovereignty*. Maryknoll, NY: Orbis, 2008.
United Nations. "Antarctic Treaty." www.ats.aq/documents/ats/treaty-original.pdf.

Select Bibliography

———. "The Declaration on the Rights of Indigenous Peoples—With an Introduction for Indigenous Leaders in the United States." Tucson: University of Arizona Indigenous Peoples Law and Policy Program, 2012.

———. "Millennium Development Goals." 2000. http://www.un.org/millenniumgoals/.

———. "Universal Declaration of Human Rights." http://www.un.org/en/documents/udhr/index.shtml.

———. "United Nations Treaties and Principles on Outer Space." New York: United Nations, 2002.

———. "World Charter for Nature." http://www.un.org/documents/ga/res/37/a37/r007.htm.

Vallee, Jacques. *Dimensions: A Casebook of Alien Contact*. San Antonio: Anomalist, 2008.

Wilson, Edward O. *The Creation: An Appeal to Save Life on Earth*. New York: Norton, 2006.

———. *The Social Conquest of Earth*. New York: Liveright, 2013.

Wright, Robert. "Ethics for Extraterrestrials." *The New York Times*, May 4, 2010, http://opinionator.blogs.nytimes.com/2010/05/04/the-moral-alien/.

INDEX

Abel, 51, 52, 95, 160
Abiota, 5, 7, 135, 241, 282
Alien(s), 1–4, 15, 17, 22, 28–30, 31–53, 63–70, 71–74, 88–90, 95, 98–99, 113–20, 128, 131, 134–41, 144–45, 150–52, 160–64, 168–69, 176, 178, 191, 199, 202, 204–15, 216–26, 233, 241–45, 254, 258, 262–63, 264–67, 279–80, 281, 284, 288
Anomalous, anomaly/ies,, 75, 77–78, 168, 172
Annas, George, 93–95
Anthropocentric/ally, 21, 28, 108, 157, 176, 258, 268, 282
Anthropomorphic/ally, 21, 84, 87, 176, 179, 257, 268
Astrobiology, 173, 213, 214,
Autopoeisis/etic, 200, 251, 258, 273–74, 283
Avatar, 9, 11, 17, 28, 42, 60, 70, 74, 90, 145, 206, 207, 212–34, 285
Ayala, Francisco J., 180–85, 194–96, 255–56

Berliner, Don, 114
Berry, Wendell, 136–37
Big Bang, 7, 200, 205, 238, 248, 257
Biota, 6–9, 14, 18–22, 27, 31–37, 53, 59, 63, 69, 84, 107, 108, 112, 131–33, 135–38, 141, 157–59, 166, 174, 178–79, 181–89, 195, 212–15, 222, 224, 227, 239, 241–42, 251, 254–57, 260–65, 270, 277, 281–82, 285
Bradbury, Ray, 82–85, 89–90
Butler, Smedley Darlington, 61

Cain, 51–52, 95, 160
Calendar, Maya, 11
Carey, Thomas J., 114
Churchill, Ward, 55
Churchill, Winston, 128, 171
Commonweal, 133, 150, 242, 282–83
Community, biotic, 5–7, 17, 19, 29–30, 90, 163, 234, 279–81
Contact, 1, 4, 18, 20, 23, 25–29, 44, 52–53, 65, 69–70, 72, 74, 76, 80–82, 84, 89, 95, 104, 106–8, 111, 131, 132, 134, 135, 136, 139, 140–41, 143–44, 150, 158–60, 162–65, 168, 196, 199–200, 213, 224, 232, 233–35, 241–42, 259, 261, 262, 265–66, 269, 278, 279–80, 285–86
Corso, Philip J., 98, 118–19, 140, 162
Cosmic Commons, 8, 24, 26, 28, 29, 42, 75, 89, 104, 115, 131, 143, 159, 185, 186, 187, 188, 281
Creatio ex Dei, 200
Creatio ex materia, 200
Creatio ex nihilo, 200
Creatiocentric, 281–82
Creationism, -ists, 171, 183, 190, 192, 251–53, 256, 257

Dances With Wolves, 225
Darwin, Charles, 8, 16, 19, 25, 34, 35, 110, 166, 174, 181, 182, 238, 253
Day Without Yesterday, 166, 200, 205, 238, 247, 248, 253
Deere, Phillip, 264–65
Demetrios, Archbishop, 160–61
Deride, -ision, 23, 73, 100, 173

Index

Deus ex machina, 8, 10, 11, 15, 84, 226
Deus sans machina, 10
Discovery Doctrine, 8, 17, 42, 53, 53–69, 72, 82–85, 90, 95, 111–12, 134, 144–45, 150, 151, 155, 158, 177, 204, 209, 215, 222, 226, 228, 229, 280, 284
Displacement, 107–13, 130–32, 215, 230
Distribution, 5, 29, 64, 132–33, 147, 159, 214, 285
District 9, 28, 178, 215–23, 227–34
"Dolphin Mission," 92–94
Donne, John, 286
Door, church house, 23

Encounters, close, 26, 73–81, 126, 212
Environment, 6, 14, 16, 22, 25, 29, 55, 72, 108, 131–32, 135–37, 145, 156, 157, 165, 179, 181, 191–93, 204, 214, 224, 227, 231, 232, 234, 253–54, 257, 278, 280, 287
Ethics, 5, 29, 226, 245, 282
   Lifeboat, 12–16
   Lifeship, 12–17
   Cosmosocioecological, 283–84, 286
   Cosmosocioecological praxis, 284–86
   Praxis, 282–83
   Socioecological, 17, 234, 281–82, 285, 286
Evidence, 24, 26, 28, 73, 76–80, 91, 93, 100–105, 106–9, 112, 118, 130, 132, 142, 162, 168–74, 179, 185, 187, 192, 194, 214, 215, 237–41, 244–47, 259–62, 279
Evolution, 1, 3, 11, 17, 19–20, 24, 25, 29, 34, 68, 82, 88, 89, 93, 95, 100, 110–12, 132, 137–39, 141–42, 162, 166, 174, 178–96, 200, 212, 214, 232–33, 243–46, 249–61, 267, 270, 274–75, 286
Exploiters, 136–37

Faith, 10, 14, 23, 25, 54, 87, 100 113, 170, 171, 176, 180, 184, 186, 187, 188, 191, 195–203, 210, 229, 237–38, 240, 243, 248–50, 254–55, 258–63, 270, 278, 286
Foner, Philip, 148
Foreigner, 33, 43–45, 48–49, 67, 68, 160–61
Francis of Assisi, 93, 241, 277
Friedman, Stanton T., 114
Funes, José Gabrielle, SJ, 241, 277

Galileo Galilei, 25, 109–10, 238, 239
Gap
   economic, 18
   evolutionary history, 182
   God of the, 202
   income, 18
   science-religion, 181
Garrido, Juan, 37–38
Genocide, 55, 63, 67, 69, 83, 85, 90, 228, 232
Goods, natural, 3, 4–5, 7, 9, 12, 14, 17, 22, 28, 29, 31, 33, 42, 53, 55, 56, 57, 59, 60, 61, 62, 63, 64, 67, 69, 71, 72, 83, 86, 90, 112, 134–36, 145, 153–54, 157–59, 165, 203, 204, 211, 231, 233, 255, 280, 283, 284, 285

Halt, Charles I., Colonel, 128–29
Hardin, Garrett, 12–16
Hawking, Stephen, 8–18, 27 52, 65, 69, 84, 86, 96, 108, 135–36, 141–42, 187, 225–27, 234, 264, 266, 268, 284, 287
Heaven(s), 10, 16, 18, 26, 28, 29, 37, 48, 62, 63, 64, 72, 74, 75, 93, 125, 126, 134, 144–45, 150, 152, 172, 174–76, 178, 179, 201, 203, 206, 226, 229, 232, 241, 253, 269, 280
Hellyer, Paul, 98, 173, 241, 265–66
Hill-Norton, Admiral, 130
Holocaust, American, 55
Homo sapiens, 63, 95, 108, 138, 185, 191
Huxley, Aldous, 138
Hynek, Josef Allen, 75–81, 90, 93, 99, 101, 114, 120, 121, 124, 261–62

Imbrogno, Philip J., 120,

Index

Immigrants, 1, 3, 15, 23, 32, 37, 40, 41, 42, 55, 65–67, 83, 86, 161, 210, 220, 230
Intelligent Design (ID), 180–83, 184, 192, 196, 238, 251–56
Invisible hand, 12, 146–49, 204

Jenkins, Jerry B., 10–11, 16, 25
Job, 44–47
Johnson v. M'Intosh, 58–59
Jonah, 44–48, 94

Kean, Leslie, 69, 79, 113, 120, 129, 130

LaHaye, Tim, 10–11, 16, 25
Lakota, 8, 210, 222, 225, 265, 277
Lemaître, George, 166, 200, 238–39, 242, 246–49, 256
Logos, 7, 200, 270–272, 274–77
Logoi, 7, 200, 275
López de Gómara, Francisco, 38

Machina ex Dei, 252
Malmstrom Air Force Base, 96–97, 118
Manna, 144–45
Marcel, Jesse, Jr., v, 98, 114–20
Marcel, Jesse, Sr., 114–20
Marcel, Linda, 114, 117
Mariana, Nick, 99–100
Martian Chronicles, 82–90, 94
Material/ly, 7, 8, 21, 23, 24, 50, 54, 57, 81, 85, 87, 88, 102, 117, 135, 136, 141, 154, 157, 159, 171, 170, 174–75, 176, 179, 182, 187, 188, 189, 190, 191, 194, 198, 199, 200, 201, 202, 204, 205, 208, 236, 237, 238, 240, 242, 256, 257, 259, 261, 262, 274, 278
Materialism, 190, 236, 237
Materialist/ic, 172, 191, 192, 199, 200, 201, 202, 237, 240, 249, 255, 257, 263
Materiality, 7, 24, 110, 113, 179, 188, 189–90, 192, 194, 198, 201, 202, 205, 236, 237, 239–42, 252–53, 255, 259, 261, 262, 268
   Sacred, sacral, 240
Mathematica ex Dei, 252

Maximus, Saint, 7, 200, 275
Metaphysical/ity, 77, 84, 138, 176, 177, 190, 192, 199, 203, 238, 248, 256, 257, 262, 278
Miller, Robert J., 53
Minuteman Missiles, 96ff
"Mission, Dolphin," 92–94
Mitakuye oyasin, 8, 265, 282
Mondragón, 13
Muskogee, 264–65

Namasté, 7, 8, 277, 282
NASA Astrobiology Institute, 214–15
National Aeronautics and Space Administration (NASA), 19, 82, 124, 151, 213, 214
National Oceanic and Atmospheric Administration (NOAA), 36
Newcomb, Steven, 55, 58
Nurture/r, 6, 95, 137

Omega point, 89, 269, 274–76

Paine, Thomas, 146–51, 156, 204, 280
Peacocke, Arthur, 182, 242, 249–51, 267, 270–276
Peking Man (*Sinanthropus pekinensis*), 89, 244–45, 249
Philoxenia, 48, 160–61
Pratt, Bob, 121,
Primeval atom, 238, 247–49
Prayer, Lord's (Our Father), 144–45

Qur'an, Holy, 52

Racism, 18, 41, 60, 66, 221
   Cosmo, 177–78
   Eco-, 231–33
Ramey, George, General, 115–16, 117
Relational/ity, 7, 8, 20, 57, 107, 234
Relational religious, 240,
Resource, 4–5, 9, 59, 62, 72, 81, 153, 159, 218, 222
Ridicule, 2, 24, 74, 76, 77, 78, 81, 100, 103, 104, 114, 119, 121, 122, 132, 168, 170, 173, 238, 247, 263
Rights, natural, 135, 148, 163

Index

Ruth, 44–45, 49
Sagan, Carl, 82, 213
Salas, Robert, 97, 131–32, 284
Samaritan, 44, 47–48, 50, 52
Schmitt, Donald R. 114
Scientific American, 2
Search for ExtraTerrestrial Intelligence (SETI), v, 52, 186, 213–14, 225, 238
Sin, original, 20–21, 87, 244, 269
Singer, Peter, 141–42, 264
Smith, Adam, 12, 13, 146–49, 150, 156, 204, 280
Sociality, 7, 142
Socioecological, 17, 27, 30, 53, 62, 83, 89, 157, 234, 281–83, 286–87
Spiritual/ly, 6, 7, 11, 18, 20, 21, 24, 26, 28, 43, 52, 54, 55–59, 72, 88, 93–95, 111, 132, 153, 164–65, 174–77, 179, 180, 181, 187, 188, 189, 242,
Spirituality, 6, 7, 24, 68, 88, 177, 180, 188, 189, 241
Stannard, David E., 55
Star People, 263–64, 265–66
Stranger, 43–44, 48, 50–52, 68, 160–61, 199

Teilhard de Chardin, Pierre, 89, 242–46, 249, 267–69, 271, 274–75
Terra, 21–22, 56, 59, 63, 71–72, 83, 86, 96, 107, 109, 135, 144–45, 208, 282
Terraform, 83, 86

Treaty, Outer Space, 8, 62, 111–12, 141, 154, 158, 204, 280, 285

Unidentified Aerial Phenomena/on (UAP), 75, 81, 92, 120, 167, 169
United Nations documents, 62, 151–54

Vallee, Jacques, 90–91, 93
Value
  Instrumental, 5, 35, 57–58, 69, 135, 138
  Intrinsic, 5, 35, 57–58, 69, 135, 138

War of the Worlds, 74, 128, 207, 211, 212
Warming, global, 8, 36
Welles, Orson, 74, 128, 207
Wells, H. G., 74, 128, 207
Wilson, Edward O., 25, 180, 184–93, 194, 196, 201, 203
  The Creation, 25, 184, 194, 195
  The Social Conquest of Earth, 184, 185, 186, 187, 188–92
"Woodstock," 7
Wright, Robert, 141–42, 264

Xenophilia, 28, 48, 68, 134, 144, 145, 160, 161, 162, 165, 280
Xenophobia, 28, 45, 67, 134, 144, 145, 150, 160, 162, 165, 232, 280

Year, Jubilee, 43, 148

www.ingramcontent.com/pod-product-compliance
Lightning Source LLC
Chambersburg PA
CBHW032051220426
43664CB00008B/958